T0130750

Cloning after Dolly

Cloning after Dolly

Who's Still Afraid?

GREGORY E. PENCE

ROWMAN & LITTLEFIELD PUBLISHERS, INC.
Lanham • Boulder • New York • Toronto • Oxford

ROWMAN & LITTLEFIELD PUBLISHERS, INC.

Published in the United States of America
by Rowman & Littlefield Publishers, Inc.
A wholly owned subsidiary of The Rowman & Littlefield Publishing Group, Inc.
4501 Forbes Boulevard, Suite 200, Lanham, Maryland 20706
www.rowmanlittlefield.com

PO Box 317
Oxford
OX2 9RU, UK

Distributed by National Book Network

British Library Cataloguing in Publication Information Available

Library of Congress Cataloging-in-Publication Data

Pence, Gregory E.
Cloning after Dolly : who's still afraid? / Gregory E. Pence.
p. cm.
Includes bibliographical references and index.
ISBN 0-7425-3408-1 (hardcover : alk. paper)
1. Cloning—Moral and ethical aspects. I. Title.
QH442.2.P458 2004
176—dc22

2004004742

Printed in the United States of America

∞™ The paper used in this publication meets the minimum requirements of American National Standard for Information Sciences—Permanence of Paper for Printed Library Materials, ANSI/NISO Z39.48-1992.

Jim Rachels's premature death at age sixty-two leaves us all worse off. Those around him lost a teacher, mentor, singer, thinker, film critic, and good friend. The world also lost a pioneering bioethicist and lover of all things intellectual, especially philosophy. I miss you, Jim.

Contents

Preface: The Political Theater of Cloning

At my congressional hearing, I didn't expect Raël, his hair in its characteristic topknot above his famous white silk robe. Back in March 2001, I thought a congressional hearing would be dignified; back then, I didn't understand how members of Congress craved media exposure. As he would prove again and again, Raël knew how to give them that.

Perhaps I should have known this, as I walked past the elaborate security checks to enter the Rayburn House Office Building, hurried down the long corridor to the huge hearing room used during the Watergate hearings, walked past many small rooms that contained hearings where one lonely member of Congress presided over a hearing with one lobbyist on topics far more important than cloning, such as the disposal of nuclear waste.

Such meager attendance at these other hearings did not prepare me for what I saw as I entered the Sam Ervin room. Reporters filled hundreds of chairs, while television cameras, including a robotic C-SPAN 2 camera beneath the dais, filmed every word. A dozen members of Congress listened and questioned, including the stars, Rep. James Greenwood (R-PA) and Rep. Diana DeGette (D-CO); behind them dozens of aides flitted about, whispering in the representatives' ears and handing them messages.

Long before all this began, Alan Slobodin, Rep. James Greenwood's staff lawyer, had negotiated who would be there, how long each participant would talk, and where each would sit. There would be no surprising the members of Congress: each of us had to deliver fifty copies of our prepared witness testimony to the Capitol a week in advance.

I had argued passionately with Slobodin, who masterminded the entire hearing, that he should not invite Raël, Brigitte Boisselier, or Panayiotis Zavos. "They're all perpetuating a hoax," I insisted. "Allowing them to speak will only give them credibility." Little did I know what Slobodin knew then: that without those controversial participants, few reporters would have covered the hearing. Baited by the eccentrics, assignment editors decided to fish. And because ordinary people did not know that Dr. Zavos was not an M.D. but a Ph.D. (whose dissertation discussed turkey sperm), they thought him a real physician about to clone a baby. Brigitte Boisselier, who has a Ph.D. in biochemistry, worked in a field that many people (incorrectly) thought might give her some expertise about cloning. This was no hearing; it was a show.

I sat at a table with several panelists, including University of Pennsylvania bioethicist Art Caplan on my right at one end of the table and, at the far left end, Raël, Boisselier, and Zavos. In between sat theologian Nigel Cameron, gay-rights activist turned cloning activist Randolfe Wicker, lawyer and advocate for the infertile Mark Eibert, and some others. Behind us sat the deputy director of the Food and Drug Administration (FDA), who later testified to open skepticism that the FDA could regulate cloning (the FDA only has the legal power to regulate drugs and devices, so by what authority it would regulate cloning is unclear).

This hearing demonstrated society's bewildered reaction to the prospect of human cloning, which includes reasoning, political theater, faith-based power politics (right-to-life activist Richard Doerflinger huddled behind me with his crowd), industrial lobbying (well-dressed biotech lobbyists glad-handed people in the crowd, wanting to protect their companies' rights to perform research involving stem cells and embryonic cloning), and the media, pushing its sensationalistic cloning stories to sell newspapers.

In writing about these contexts, I would be dishonest if I discussed only the rational and scientific arguments about cloning because much of the controversy, wildness, and emotion would be left out, which—for better or worse—shoots bolts of lightning into the continuing story of cloning.

Cloning is the most important bioethics issue of our decade. President George W. Bush talked about cloning stem cells in his first prime-time television talk to the nation, which was also the first time a U.S. president ever addressed the nation about a topic in bioethics. As cloning has become the Incredible Hulk of national bioethics public policy, others have come to the table: ministers, lobbyists, scientists, the United Nations, and trade representatives. It is as if a seer prophesied to the newly cloned baby, "May you live in interesting times."

The relationship between cloning and the media matters a lot here. During my congressional hearing, aides and members of Congress flipped through copies of the February 2001 issue of *Wired* magazine, which featured a sensationalistic cover story ("Me—Again?") that allegedly exposed would-be parents and covert scientists working to clone children. This shocked me because I had spent many hours helping the freelance writer with this story and I had disputed these claims. The writer's final sensationalistic result—designed to sell millions of newsstand copies—disgusted me. I knew that no such labs existed, that no American physicians were anywhere near to cloning a child, and that no reclusive billionaires were trying to clone themselves in a repeat of David Rorvik's 1978 hoax *In His Image: The Cloning of a Man.*

During the preceding year, the *Wired* story and the congressional hearing had been building to a crescendo, orchestrated behind the scenes in large part by maestros of the national media, collaboration by a half dozen eccentrics who could have come from central casting (a Chicago scientist who looked like a homeless man and was called "Dick Seed") and by faith-based groups who, blindsided by abortion and failing to overturn *Roe v. Wade,* now rallied around the issue of cloning as they prepared to storm the congressional barricades.

Stephen Toulmin once wrote a famous piece, "How Bioethics Saved Philosophy," describing how in the late 1970s applied ethics revitalized the moribund field of philosophical ethics that had been stuck for a decade in technical issues of meta-ethics. Today one could write "How Cloning Saved the Religious Right" by giving it a new issue besides abortion. Each time Zavos, Italian physician Severino Antinori, or Raël goes on television, religious conservatives scream that the biotech Armageddon is at hand (my favorite alarm is a video, *Cloning to Kill,* by television evangelist Rod Parsley that shows small babies in test tubes with voice-overs about scientists who want to kill them). As Greenpeace needed genetically modified organisms (GMOs), fundamentalists needed cloning.

While the freelance writer working for *Wired* sought secret cloning labs and people ready to clone themselves, researchers from major television news shows, national magazines, and national newspapers were doing the same. During March 2001, not a week went by when I didn't get a query from some reporter earnestly looking for such labs or "couples who wanted to clone their dead child."

Time went with a fast cover story to compensate for its previous week's story on African famine. ("They lose money on down stories," one freelance writer told me. "No one buys newsstand copies with depressing covers. So the sensational cover story on cloning makes up for it.") If you watch, you begin to see that any commercial news medium with a depressing story must balance it with an exciting story, not only for psychological balance but also to pay the bills. This is true for everything from the *New York Review of Books* to *60 Minutes.* Stories on cloning pay for stories on the plight of the homeless.

Likewise, no one comes to congressional hearings that are boring. Not many television reporters would have come to hear scientists hold forth about cloning. What I didn't know then, but what Representative Greenwood and Alan Slobodin knew, was that inviting Raël, Brigitte Boisselier, and Panayiotis Zavos would guarantee saturation coverage by all the national print and visual media. And if it gave credence to kooks, who cared? The point is to get one's face and issue before the public, and the attracting of the media is the key to doing that.

Two years later in March 2003 I talked about cloning at a northern university. This was a few months after the onslaught of news coverage during Christmas week 2002 about claims to have cloned a baby by Brigitte Boisselier, Raël, and their company, Clonaid.

College students changed radically during this pivotal week; before, few had been openly hostile, afterward, the majority was. It was not abnormalities about cloned embryos that changed their view, but the association with eccentric people who had foreign accents and wore funny costumes and believed that extraterrestrials created humans. The damage to rational arguments for reproductive cloning done by the Raelians exceeded what Jack Kevorkian did for physician-assisted dying. They turned a serious topic into a joke, and consequently in this book I face the almost impossible obstacle of arguing for something that everyone has already decided against.

Cloning fascinates everyone. On a cable television show, MTV's *Clone High USA*, genotypes have been resurrected of John F. Kennedy, Cleopatra, Abe Lincoln, and Joan of Arc. Right now, cloning cannot be extricated from politics, so I'll discuss the politics of it in this book.

When safe and efficient, cloning will alter humanity less than science fiction predicts, but in other ways, more than people understand. And now to explain just that.

A NOTE ON THE IMPORTANCE OF NONPREJUDICIAL WORDS

Throughout history, humans have used language to express judgments of value. Unfortunately, and because language is so rich and subtle, some of those words express sharp negative judgments about whole classes of our fellow humans. Scholars of language know that a word is much more than its literal meaning. It possesses connotation, denotation, illocutionary force, an etymology, strong associations, sometimes a political pedigree, and different usages in different contexts.

Most of us also know that bigots use words as weapons, words such as "nigger," "fag," "chick," "prick," "untouchable" (*bhangi* in Hindi), "wop," "queer," "bitch," and "Jew." The list goes on. Sometimes, however, an entire culture is bigoted and uses such terms unaware of its own universal prejudice until its bigotry is challenged by reformers. Two generations ago, whites in the South referred unself-consciously to "darkies" and before that, in another phrase that makes us shudder, President Theodore Roosevelt called for expelling "yellow niggers" back to Asia.

Yes, words can hurt and the way they are used can beg important ethical questions. For these reasons, I will not use the word "clone" or the phrase "the clones" in this book to refer to humans originated by asexual reproductive cloning. "Clone" by now connotes aberrant, mass-produced, commodified subhumans. It is a deeply prejudicial term, as if the authors at a law school on the Equal Rights Amendment began their report, "The chicks here at the law school say women are the equals of men. . . ." As soon as you read the first two words, you would know the conclusions.

I continue to be amazed at how advocates for human cloning use these prejudicial terms, allowing the negative connotations to put them at a disadvantage in discussions. Critics are adept at manipulating such prejudicial language, asking seemingly straightforward questions such as, Would the clone have a soul? Would a clone of Hitler be like Hitler? Would clones of Michael Jordan . . . ? In all these phrases, the emotional punch of "clone" and "clones" is decidedly negative.

So I will slog through with "person created by cloning" or "cloned baby," and beg the reader's forgiveness for my sometimes awkward language. Over time, I hope, we will all gravitate to simpler terms, such as "person."

Acknowledgments

No author writes a book without some kind of help and emotional support. The first person I acknowledge has been my Number One fan since my birth; at eighty years of age, she has become a one-woman clipping service on every topic under the sun relating to biotechnology. Every other day, a packet arrives in the mail, and frequently something in the packet is news to me. Thanks, Mom!

At a crucial time in my writing, when my energy was flagging, San Francisco attorney Mark Eibert read what I had written and urged me on to higher and better places. A remarkable lawyer and an advocate for infertile people, Mark is a brilliant person who really thinks for himself. I am lucky to count him as a friend.

From the other side of the country in New York City, cloning activist Randolfe Wicker also continued to send me items and keep me abuzz of everything happening in the world of cloning. Thanks, Randy, and I hope to read your book one day soon.

My colleagues N. Scott Arnold, Ted Benditt, Lynn Stephens, Harold Kincaid, and Maureen Kelley either read key chapters or helped me with good conversations. As ever, I owe a special debt to the extraordinary help of Mrs. Minnie Randle. I also want to thank copy editor Chrisona Schmidt, production editor Julie Kirsch, and Rowman & Littlefield acquisitions editor Eve DeVaro.

My wife, Pat Rippetoe, continues to supply me with tips about cloning in the media and spotted the original Discovery Channel show on humanzees. She patiently endures my obsessions about cloning and biotechnology.

Jim Rachels was also a key person with whom I discussed ideas in this book. He contributed in many ways that are not obvious.

Parts of this book were developed from talks at the Transhumanist Society meeting at Yale University in June 2003, the conference on the ethics of the artificial womb at Tulsa in February 2002, the Fourth National Undergraduate Bioethics Conference at Notre Dame in 2001, the Fifth National Undergraduate Bioethics Conference at Emory University in 2002, the Sixth National Undergraduate Bioethics Conference at Texas A&M University in 2003, the conference on cloning sponsored by the Department of Neurobiology at Northwestern University in May 2003, and the European conference on cloning and human nature at Lausanne, Switzerland, in February 2004. I also acknowledge the superb proofreading by my 2004 summer interns, Anand Iyer, Roshan Patel, Emily Taylor, and Qin Zhang.

1

How Cloning Will Surprise Us

ADRIANA: I see two husbands, or mine eyes deceive me.
DUKE: One of these men is genius to the other.
And so, of these, which is the natural man?
And which the spirit? Who deciphers them?

—*William Shakespeare,* The Comedy of Errors *5.1*

Cloning will surprise us, and I'm not talking about the fact that cloned people will differ from their genetic ancestors. Harvard biologist Richard Lewontin, Oxford biologist Richard Dawkins, and a previous book I wrote all emphasized that our adult selves also stem from environmental influences and that no one should accept genetic reductionism.[1] No, we should no longer beat that dead horse.

Nor am I thinking about the fact that mitochondrial genes in the host egg would contribute a tiny number of genes to the new person, and that these genes would make the resulting person different from the ancestor. Nor am I contemplating the fact that in sexual reproduction, a certain number of genes on the X chromosome randomly inactivate in forming an embryo, and thus identical twins differ, as embryos, regarding which genes are active, which are not. Tiny differences at early stages account for later differences between identical twins. But this is standard twin lore. No whistle here.

And literature? The basic theme of all literature about biotechnology goes back to the mother lode, Mary Shelley's *Frankenstein*, and her alarmist theme that arrogant scientists who mess with nature, thinking they can control, improve, or contain it, will be surprised by the results.

Yesterday's writers speculated about the tumultuous relation between a child and his clonic ancestor. In "The Extra," by Greg Egan, retarded cloned adults are created as sources of extra organs for the ancestor, but they get their revenge. In "Phantom of the Plains," a cloned woman tries to kill her ancestor, a composer, because she can't be as talented as the ancestor. And there is the perennial tale (e.g., "The Cloning of Joanna May," by Fay Weldon) of the girl cloned to be the young version of the rich old man's wife, who—surprise!—finds the old man repulsive and runs off with a young stud.

Other stories reveal that the person cloned resents the motives of the ancestor who originated him and won't cooperate with the ancestor's plans. The cloned person resents the ancestor's expectations and refuses to be an athlete, actor, rock star, or scientist. Is this new? Interesting?

On rare occasions, cloning literature shocks us. In Lisa Tuttle's naughty little story, a handsome, lusty young man who was cloned and raised apart from his male ancestor meets his ancestor one day as one male stranger meeting another.[2] The shock is that the two adults, both gay and oversexed, immediately lust for each other and get it on. In all these stories, we find the familiar ring of a folktale moral: Don't mess with Mother Nature or she'll burn you; don't bait the gods or they'll strike you down for hubris; don't design children or they'll turn out to be freaks.

The record of futurologists in regard to medical science is dismal. Many of them predicted horrors that never came about, while on the other hand, some of the most revolutionary developments were never foretold. Alvin Toffler predicted in *Future Shock* (1970) that we would send embryos into outer space, where they would gestate, be raised in robotic nurseries, and then become space explorers. Instead of this fantastic scenario, today our space program hovers on the brink of disintegration, as the future of the space shuttle is increasingly uncertain.

Today we fear eugenic marriages and sterilizations by the state, remembering the injustices of the Nazis and our little-taught history of our own eugenic policies in North America in the twentieth century.[3] *Gattaca*, *The Handmaid's Tale*, and *Oryx and Crake* led us to expect that genetic knowledge will create dictatorial genetics. But the real lesson of the early eugenics movement is that what is thought of as "normal" is what later bites you.

What you do *not* see turns out to be the deep moral issue, what later generations take to be your stupidity, racism, or unstated assumptions.

Clearly making predictions about the future of cloning and popping the balloon in advance of surprises is risky. But what might the future hold of a different sort for human cloning, once it has been made safe? Perhaps the ancestors of cloned people might learn lessons they didn't want to learn.

The cloned person created from your genes might teach you how, but for tiny slips of fate, you might have been great. For example, suppose you are smart and strong, but suppose that your mother, who bore you after your father was discharged from the army following World War II, drank during your gestation (as most mothers did) and never took folic acid (as most mothers did not). From your cloned child, you realize how greatly your mother's small sins affected you, as your clonal descendant has a tested IQ of 150 rather than 130 and has the kind of confident curiosity you wish you had.

One day a very bright white student told me a story. The student had grown up in rural Alabama, and even though his grades were very good and his test scores in the top 1 percent nationwide, neither he nor his parents ever thought of applying for a scholarship. No one in his family had ever graduated from high school, much less college. That he would deserve such an award or actually get it was beyond their cognitive landscape. So he went to a public commuter university, the University of Alabama at Birmingham (UAB); only after excelling there, did he begin to think of winning support for further study.

So the ancestor may realize, after his cloned child receives scholarship offers, what he might have been offered if his parents had been more savvy about scholarships or about getting along with classmates and members of the opposite sex, about the overemphasis on athletics in schools to the detriment of the intellectual life, and about the general agony of teenage years. Who doesn't wish he had a wise guide through those years? Someone might say to her, "Don't let people tell you your acne is just a passing teenage thing. If you don't get treatment, your acne will produce deep scars. See a dermatologist now!"

Or suppose you never had a close bond with your father, never had a father who enthusiastically supported your plans to play classical music and the cello. Not wanting to make the same mistake, you wholeheartedly support your child, seeing him flourish in countless directions. But you also sadly realize that with a more supportive father, you could have been great.

And that is just the beginning of what our cloned progeny might teach us. "If only you had tried harder, Dad, you could've published your book on James Joyce. You had the ability! After all, I published ten books before I was forty and you had a whole lifetime!"

Multiple copies of an ancestor's genes might bring multiple surprises. Having multiple copies of the ancestor's genome could create multiple chances to see the limitations of the ancestor's efforts, drive, attitude, or choices. When critics say human cloning is about control, always remember that the free will of cloned human beings cannot be controlled by their genetic ancestors.

In another direction, having multiple copies of a genotype would be a naturally controlled experiment, whether designed that way or not. The genome of the ancestor is the control, and variations in genes, environment, or choice will show how things could have been different. Such experiments might reveal painful truths to the ancestor, such as how small improvements in nutrition or vaccination can result in an extra fifty years of vigorous life. Along the way, of course, such natural controls and variations will teach the rest of us important truths.

Another surprise might be this: suppose some aspects of genetic reductionism are true but not those we expect. As the Princeton poet C. K. Williams speculates, suppose my morbid fear of death is in my genome? Won't the ancestor be blamed for creating another person with a similar life-crushing fear? "The first thing I imagine with him is my fear. To see oneself naked before one's naked self. The human dream, the human dread."[4]

For such a fear-based person, Williams continues, "I become afraid that he'd be *more* than me, just as much as ever me, but more focused, less susceptible to all the flinchings and feintings that are the residue of all I've lived. With greater *ardor*, he exists, with *greater force*." (But if my cloned child didn't have my "flinchings and feintings," my fears, my dread—he wouldn't really be me. What if those fears are close to my essence? Can a fear-based person be the same person without his fear? Maybe not, if Peter Kramer is right in *Listening to Prozac*.)[5]

Williams also fears his adult cloned child might be too good, an even better poet, and replace him. At first people might ask, "Was that C. K.

Williams the first or the second?" But as the years go by, they would blend together, and if the second was clearly superior, only the second would be remembered, just as James Mill is only remembered for creating and educating John Stuart. So I speculate, wondering if my adult cloned child would write such a book better, having attended better schools and having received more encouragement. Perhaps I will leave him this house on this river beside this small mountain in Alabama, so he can write in the same tranquil setting. But if he turned out to be better than I in all things that I value, I do not begrudge this to him. (Just let him remember his ancestor.)

We may also be surprised that cloned children experience their chosen genotype as a blessing, not a burden. In my experience counseling undergraduates, most of whom come to the university with no fixed idea of a future career or job, I hear statements like, "I wish there was something special in store for me. I wish I was *made* for something!" Indeed, the whole quest to "discover who I am" assumes there is something to discover, not simply choose. If it is a choice, many young adults find such a choice paralyzing: "What if I choose wrong? My life will be wasted! How do I know how to choose? What criteria should I use?"

Consider a thought experiment. Suppose that a college student was created from the genotype of a great architect. She was good at drawing and always built things in her parents' tool shop; wary of the burden of expectations, her parents never told her of her genetic ancestor. For some young adults, it would come as a relief to learn of this ancestor. There would be a rush of meaning in their existence: "Ah! Now I understand why I want to do certain things and why they feel right!"

Of course this is a chancy prediction, but it is made against the agony many undergraduates face of what could be called a criteria-less choice: many drift into a career rather than choose one, sometimes with the shaping of parents, often arbitrarily. ("A job opened, I applied for it, and got it. That was the beginning.") Many young people suffer from lack of direction and lack of confidence in future success; they hunger for a special calling. Coming from the genes of a successful ancestor might give them that meaning in life. People have found such meaning in much stranger places. (For example, in belonging to a family tree in which one only shares half the genes of the parent with the name of the family, or a quarter of the genes with the namesake grandfather of the family, or one-sixteenth with the namesake great-grandfather.)

A different kind of surprise is a political surprise. In the debate over nature versus nurture, a certain kind of strategizing can be detected by egalitarians and antiegalitarians. Egalitarians fear evidence that differences in achievement among humans depend on innate differences, not individual effort and schooling. Antiegalitarians welcome the failure of programs such as Head Start because they believe they are a waste of money and do not affect later achievement.

Antiegalitarians these days are reductionists, asserting that high achievers come from good families with good genes. In contrast, egalitarians resist evidence that genes ground abilities and hope that environments can largely shape abilities. They hope children are largely a blank slate, so that loving parents, a great school, and a social support network can turn every child into a winner. They fantasize that more money or a more nurturing family can change environments of poor children for the better; hence, social progressives are educational environmentalists. In contrast, if it's truly all in the genes, efforts to promote equality through better environments are wasted.

With this ideological battle in mind, one particular threat from safe, regular human cloning looms large. The threat goes like this: it is one thing, egalitarians assert, to tolerate inequities in support of families, early childhood education, and quality of K–12 school systems (although liberal democracies in fact tolerate this kind of unfairness by allowing children to inherit wealth and by allowing private schools). Beyond that, however, the prospect of safe cloning raises the possibility of a categorically different kind of inequality hitherto unknown among humans: an inequality written not in the shifting sands of unequal environments but in the cold stone of genes. Liberals might draw the predictable conclusion that although environmental inequality is tolerable in liberal democracies, *biological inequality* is not. Because safe human cloning would allow biological dynasties, it should be made illegal by all countries now.

Hence the egalitarian's nightmare is a future society of billions of "Naturals" propagating just as their ancestors did for millions of years, their children illustrating the famous regression to the mean. But within that society, and increasingly leaving Naturals behind, are the "GenRich," children cloned from superior family genotypes. GenRich families that carefully control the reproduction of their children, mimicking the aristocracy of old where marriage and children concerned not love, romance, or sex but property rights and secular power.

Princeton genetics professor Lee Silver speculated about this possibility in his *Re-Making Eden*, musing that if this process were carried out a thousand generations, a new kind of species might emerge, so different from normal humans that the two could not interbreed.[6] Do we really need to worry about the possibility of a genetic overclass? Is the egalitarian objection a cogent one? I think it is, but it needs to be carefully analyzed. An old dictum in ethics says, "Ought implies can," meaning that we needn't worry about whether we ought to do something if we in fact cannot do it. If cloning can't create biological dynasties, we need not ban it.

As I shall argue in chapter 16, some elite families might use cloning as a selective tool to advance their own ends. Because families have always sacrificed for a better future for their children and created families to have something of themselves carry on after their death, the use of cloning in elite families deserves further scrutiny.

GENETIC DETERMINISM?

Behind the nature-versus-nurture debate argued by social liberals and conservatives lies the power (and it is indeed power) of the idea of genetic determinism or the perhaps more neutral term *genetic essentialism*. Demarcating the multiple dimensions of this power is difficult, but one way is to point out that genes are only part of what determines who we are; at present, however, we can't fight against or change our genetic legacy. Genetic essentialism threatens our sense of free will, an open future, autonomy, and political individualism. As Robert Wachbroit of the University of Maryland's Institute for Philosophy and Public Policy writes, "It is difficult to gauge the extent to which repugnance toward cloning rests on a belief in genetic determinism."[7]

Reproductive cloning promises to surprise us because, one way or the other, it must tell us the extent to which genetic essentialism is false, partially true, completely true, or, most likely, true in subtle but complex ways that we do not anticipate. Such bitterly contested scientific debates are usually settled over decades if not centuries by accumulating scientific evidence—one thinks of evolution, Galileo, and whether embryology shows ensoulment—but with cloning, the results may be known within two decades of the births of the first cloned children. By the time the first cloned children are in the their early twenties, we will have a good idea to what extent, if any, some traits are directly hereditable, such as musical talent, gymnastic ability, and kicking goals in soccer. Because the talents will be especially visible on the stage or in the athletic agon, the effect on general culture will be much swifter and more intense than resolution of previous scientific controversies.

If the genetic twin-child of Bill Gates excels at creating software (say, MiniSoft), it's all over. Not a thousand bioethics commissions or a million sermons will deter the enthusiasm of parents who want to give their children great advantages in the next generation, just as all good parents have always wanted. These parents should properly be called reproductive pioneers, analogous to our forefathers and foremothers who disembarked from ships in Philadelphia and traveled down the Shenandoah Valley, creating farmlands by clearing land of rocks and making labor-intensive log fences from felled trees. But these new pioneers will be in biology and they likely will have just as challenging a job.

Interestingly, the more genetic essentialism is true, the more certain arguments against reproductive cloning gain in power. The right to an open future and the right to a unique identity, both used against such cloning and allegedly repudiated by critics who claim genetic determinism is false, become more persuasive the more genetic essentialism is true. Radical biologists such as Harvard's Richard Lewontin dismiss objections to human cloning based on genetic essentialism and hence dismiss arguments assuming that cloning the genes of great people will result in, if not great people, at least great genomes.[8] But if genetic essentialism is at least partially true, then such fears will merit more than a cavalier dismissal.

SURPRISES TO THE CLONED ADULT

Demonstrating the falsity of genetic determinism will have many twists, including some for the people created by cloning. A child or adult may think she will be a great guitarist or painter but may be surprised that she lacks the innate talent. Like her parents and ordinary people, the adolescent girl may buy into (what we may learn to be) myths of genetic determinism. In particular, differences in creating the embryo, uterine gestation, or a lifetime of development in a different time and place from the ancestor's may be the difference between above average talent and exceptional genius. We just don't know. But certainly some cloned adults themselves will be disappointed too, with their own false expectations.

Is this a terrible thing? I don't think so. Once it's safe and around, everyone will be on a learning curve about cloning, and it will take some time to figure out what is true and what is not. The same can be said for every great scientific advance.

CONCLUSION

By now we know something about the lore of cloning, and certain surprises from cloning are no longer surprises. On the other hand, many other surprises will occur and depend on the extent to which genes determine our traits and how predictably. Because such claims are so steeped in politics, it is almost impossible to know how things will turn out. Preliminary evidence will be filtered through political lenses, and only when evidence is overwhelming will critics fall silent. In the meantime, we must await both safe cloning and its implications.

2

What Cloning Tells Us about Ourselves

To hold as t'were the mirror up to nature.

—*William Shakespeare,* Hamlet *3.2*

When people heard that Dolly the sheep had been cloned and inferred the possible cloning of humans, many felt the moral ground had shifted beneath them and said, "Yuck." Leon Kass, the biochemist who heads President Bush's Council on Bioethics, glorified this reaction, using his eloquence to try to elevate it to wisdom.

People fear cloning humans for many reasons. These fears tell us many things about ourselves. Many people feel that the clone would be treated badly, as a slave, zombie, conditioned killer-soldier, harem girl, or source of organ parts. Such fears center around the assumption that the cloned being would be a subperson or subhuman. The same fear manifests itself theologically in the fear that the being would lack a soul.

People fear the motives of the people originating cloned babies. They fear people who want a copy of their genes to be continued in another human. They think this is selfish and narcissistic, an insult to God's wisdom that would result in the ancestor imposing his will on the child.

People fear that cloning humans would break the mold of the old way of sexual reproduction and open a wedge leading to other forms of asexual reproduction. Critics love to cite the sensational problems and conundrums of fertility clinics with embryos—the occasional mix-ups and custody battles during divorce—that the media loves to report. Behind this particular fear lies the idea that the scientists and physicians driving this technological juggernaut are not really motivated by helping infertile people have children but by greed, arrogance, an unbalanced desire to achieve success, and a moral insensitivity that comes from being a nerd or science geek.

People worry that this new tool might benefit rich people more than ordinary people, and might not even be available to ordinary people.

Finally, there is the worry that, despite the best efforts of parents and society, cloned children would feel, when they discovered their origins, inferior in comparison to ordinary children created sexually. It is a lingering fear of being different, of being marginalized—the kind of fear reported by some gay men and recent immigrants.

It is thought that "designer children" could never live up to the lofty expectations of the parents who originated them. Children who do not want to be tennis players or physicians might nonetheless succumb to overwhelming parental expectations. Worse, children would feel they were not wanted for themselves but only for the specific traits desired by their parents.

What do these fears about cloning humans tell us about ourselves?

Cloning shines a light into the normally dark places in human relationships. Because morality is about conflicts of interests in those relationships, the morality of cloning interests us. In other words, heterosexuals don't normally talk honestly in public about their true motives for wanting children, what kind they fantasize about having, and what they've done to their own children to bring about their ideals.

News about cloning also fires our deepest fears because in condemning cloning, we project our fears. Instead of being honest about what we fear in ourselves, we impute such fears onto evil others. But like all projections, our fears tell us more about ourselves than about those who would create cloned children.

Fears about cloning do not usually relate to asexual reproduction but symbolize how we presently relate to each other and to children. As London psychoanalyst Adam Phillips says, "Cloning is, for obvious and not so obvious reasons, a compelling way of talking about what goes on between people."[1]

When people worry that clones will be sexually enslaved or made to work under inhuman conditions in desolate Ethiopian mines, they are not basing their views on evidence and reasoning but are fantasizing about what they fear or want, like the teenage boy who became fed up with the machinations of teenage girls and said, "I just want a clone girlfriend." In other words, he wants a Stepford wife, a good-looking woman who will do what he wants, when he wants. That is how he thinks about a cloned woman.

So too people would like an identical twin without a brain to be a source for organs when their own livers turn cirrhotic after years of alcohol overindulgence. Who wouldn't? Many of us would like a personal valet of the kind that English lords once had. Let's be honest here: who dreams of being an illiterate serf working all day cutting wheat? Who thinks that previously on the wheel of rebirth he was a forsaken galley slave in some ancient navy?

We feel superior when we have others at our beck and call. That is the appeal of the clone, and for many, it is a sexual appeal. How else to explain the widespread fear that a man would want to make love to the cloned daughter of his wife, if this does not "out" the same desire toward his present daughter? But because the cloned daughter has no rights, anything goes.

Or at least, that is the fear in the back of the mind which the front of the mind reacts against after repression sets in, setting up the "yuck" reaction.

And designer children? Simply a manifestation of what people want now. If people didn't see such desires in themselves and other parents, what would there be to fear? Where does such fear come from? Surely it arises when the teenage girl gets braces on her teeth, surgery to fix a crooked nose, or even breast augmentation.

Contemporary parents fear how mass culture is teaching their teenage girls to blatantly sexualize themselves: bare midriffs, skin-tight jeans, exposed cleavage. Surely their parents must feel that trying to control such powerful influences over their daughters is like spitting into the wind. And if mass culture can do this to our daughters, what will it do at the biological level, when we are talking about changing the underlying substance, not superficial characteristics?

But notice where the real fear is: parents' inability to resist the influence of mass culture on their girls. *That* is what is to be feared, not cloning itself. Cloning is only a tool that will be used by so few people—mainly the few infertile willing to sacrifice all other material needs to trying to create a child—that as a mass issue, it will be irrelevant. But people confuse what they fear in themselves, or what they fear in mass culture, with cloning.

And narcissists who want to clone themselves? That fear stems from a powerful reason why people create children in the first place: to have something of themselves to continue after death, to create a little bit of biological/genetic immortality. Or to see their own traits flourish in their children. Pride of authorship, so to speak.

But these reasons are not commonly discussed in private, never in public, so we only hear about them indirectly, when we listen with the third ear to people talking about parents who would clone themselves to create children.

And inequality. A lot of people envy what others have, and envious people don't want the world to become more unequal. But cloning is really not the real issue here, because if people really wanted to get rid of inequality, they would oppose private K–12 schools and private colleges and universities. Eliminating these and forcing all children into public education would be gigantic steps toward reducing environmental inequality. Of course, it is unrealistic to think that private schools will ever be eliminated, so people can protest symbolically by attacking cloning.

Many people feel resentful about how they are treated by others and how some other people are treated, especially those who have been historically mistreated. They fear creation of a new class of beings that may or may not be seen as persons, and who may also be mistreated.

In this sense, Ronald Nakasone (professor of Buddhist studies, University of California at Berkeley) is absolutely correct when he writes, "The cloning of human beings, like the use of artificial insemination and in vitro fertilization, is really about expanding our notion of humanity and our moral parameters."[2]

Finally, people's reactions to cloning also tell us something about how religion still influences moral intuitions. Whether that is a good thing or a bad thing depends on your view on the truth of religion's metaphysical claims, as well as your view of the usefulness of organized religion.

Consider that the most typical objections to reproductive cloning (as well as genetic enhancement) argue that "we should not play God" or "we're not God." Other criticisms bring in God and religion by invoking human nature, with it understood that God, not evolution, created us as we are.

Even objections to human cloning based on safety, unrealistic expectations, and harm to society may actually mask religious opposition. By far, the major source of all objections to human cloning is the idea that God makes babies, that God should decide how and when they come, not humans. Plus the unstated assumption that God wouldn't want humans to clone babies.

This claim is important for two reasons: (1) its explanatory power and (2) its political implications. What is the evidence that objections to human cloning are really religious? It is customarily said that the main moral objection to human cloning is physical harm to the child. Undoubtedly, great risks now exist in trying to originate a child by cloning, but why is cloning the object of federal legislation and national outcry? Thousands of medical experiments every day subject hundreds of thousands of people to risk in every developed country. If one counters that people have given informed consent to be subjects of such experiments, whereas children created by cloning have not, one can point to the hundreds of research protocols in hospitals for children where fetuses and newborns cannot consent.

Why then is cloning singled out for special objections? The answer leaps out: there is something unique about cloning that threatens the beliefs of religious people and makes them fear it. What exactly could that be? Father Richard A. McCormick of the Society of Jesus articulates this very well: "Viewed theologically, human beings, in their enchanting, irreplaceable uniqueness and with all their differences, are made in the image of God."[3] McCormick and other clergy fear that cloning is a kind of secular "manufacture" that, in the creation of the cloned baby, replaces God's hidden hand with conscious human choices.

This explains part of the power of criticizing cloning as "manufacturing" or as "commodifying" human life. The unstated comparison to God creating each human provides the foil. Similarly, kids created by God are "gifts" and are free, not something for which reproductive specialists need to be paid. No surprise that after thousands of years, our inherited ethical intuitions are steeped in God assumptions.

Even if we understand genetics and how each parent contributes twenty-three chromosomes and one sex gene to the sexually created child, and even if we understand how the genetic roulette wheel mixes the genes of mother and father to create a new, unique child, we can still believe that the hidden hand of God guides the genetic mixing.

But in regard to cloning, such belief loses credibility. Human beings choose to clone the genotype of a modern Leonardo da Vinci and not those of their neighbors, and that choice brings conscious human wants and decisions to the foreground of public policy. (Of course, if you believe God gave humans a brain to think and choose with, cloning doesn't threaten you at all, but such theists seem to be in the tiny minority and are uninterested in controlling public policy.)

A second source of religious opposition to human cloning is its link with abortion. Both cloning and abortion expand human choice over the timing, number, and traits of offspring. Many conservative religious people felt ambushed by the sudden legalization of abortion in the early 1970s and have been fighting for recriminalization ever since. Having lost this battle for thirty years, they have vowed not to lose the battle over reproductive or embryonic cloning.

The idea that opposition to cloning stems from religion is also important if we consider the basis of freedoms in a modern democracy. If the statement "Cloning is evil and should be forbidden" is defensible only by appealing to religious premises, then we have every right to ask why federal or state governments should enforce a general religious belief on nonreligious couples or on couples whose religious beliefs (e.g., Buddhism) do not incline them to reject cloning.

Perhaps aware of this problem, many religious leaders who want power in national public policy do not use religious language in discussions of cloning. Instead they appeal to vague predictions about harm to cloned children, the family, or, even more murkily, society.

Opponents of human cloning who see themselves as philosophers, such as Leon Kass, Dan Callahan, and Tom Murray of the Hastings Center, never explicitly state their religious premises but nevertheless use language that carries religious resonance. These tactics are especially insidious, for the religious source of the intuitions is never revealed while it moves their hearers emotionally.

All in all, debates about cloning humans reveal more about the humans who are debating than about future humans. When we fear sex slaves, robots, extra sources for organs, rich families, or kids who don't live up to parental expectations, we mainly are criticizing our neighbors, our own families, ourselves. If we can't accept this, debates about cloning will never be honest.

3

The Brave New World of Animal Cloning: Racing Mules and Million Dollar Cows

Animals are nothing but the forms of our virtues and vices, wandering before our eyes, the visible phantoms of our souls.

—*Victor Hugo,* Les Misérables *5.5*

For some, animal cloning is a sideshow to the main act of human cloning, a little warm-up act about rates of success and failure that may or may not generalize to cloning humans, the kind of thing that people in the agriculture business think about in terms of livestock, certainly not something relevant to humans. After all, experiments you can do on animals differ greatly from what you can do on humans. The two kinds of beings live in two different moral universes.

People who think this way are mistaken: animal cloning is paving the way to human cloning. Understanding the basics of mammalian genetics, physiology, and development is building the foundation for human cloning, just as countless animal studies have preceded human trials in medicine. Like it or not, humans are animals, and their genetics, physiology, and reproduction follow the patterns of other mammals. What we learn in general about mammalian cloning will directly translate to human cloning. Nevertheless, people think very differently about animal cloning and human cloning. Cloning animals such as pigs and rats is a curiosity, almost a freak show, not an important moral event; nothing like cloning a human child.

I think that view is naive. Biotechnology is pressuring the rigid lines between all kinds of species, adding a fish gene to a strawberry plant to prevent damage from frost and growing human embryos in cow eggs. Ignoring genetic incursions of human genes into nonhuman animals, as well as incursions going the other way, is to ignore one of the most important ethical topics of our day.

The first big picture is that the only good ethical objection to trying to originate a child by cloning is that it may produce a physically damaged child. But this objection is based on what we know at this early stage about animal cloning. As those facts change, the strength of the objection proportionally weakens. Eventually, if scientists can safely originate every other mammal by cloning, this will be good inductive evidence that human cloning is safe to try.

The second big picture concerns groups opposed to genetically modifying animals, either through cloning, knock-out genes, or genetic enhancement. Humans have modified animals by breeding them for food and pets since prehistoric times, when friendly wolves became the ancestors of today's dogs and African wildcats became the ancestors of today's cats. We have been changing animals ever since for our food, for company, and to fulfill our desires. Whether that is moral is a big question, but biotechnology itself is only doing something in a more efficient way that has been done for centuries: selectively breeding animals for specific characteristics to fulfill human purposes. Before, creating a particular kind of fish, cat, cow, or dog was a hit-or-miss process, but now the process can be much more efficient.

ANIMAL SUCCESSES

In 1995 scientists thought it was an exceptionless, unbreakable law of nature that animal cloning was impossible, but breakthroughs since the cloning of Dolly the sheep have been spectacular. Although the media harps on abnormalities and numbers of aborted fetuses, it never provides a context for these figures, for example, that one embryo has died for every human ever born, or that one in twenty humans is a twinless twin (one twin dying in utero).[1]

In an amazingly short time after the announcement of Dolly's birth, researchers cloned several different animal species. In January 1998, just ten months later, Steve Stice of Advanced Cell Technology (ACT) announced that he and University of Massachusetts scientist James Robl had cloned two calves, Charlie and George, plus six identical twins, that grew up near

Texas A&M in College Station. Stice and Robl founded ACT as a for-profit company and the identical calves were cloned to produce human serum albumin, a blood protein given to patients in hospitals. After Dolly, PPL Therapeutics created the lambs Molly and Polly to create factor IX, a blood protein that promotes blood clotting.

In June 1998, Ryuzo Yanagimachi, a biologist at the University of Hawaii, announced that he and his Team Yana had cloned three generations of mice. Yanagimachi had previously contributed to in vitro fertilization and in 2003 was one of fifteen people elected to the National Academy of Sciences. In July 2000, the Roslin Institute announced in *Nature,* while a rival team of Japanese scientists announced in *Science,* that it had cloned pigs. Roslin cloned the "five little piggies" that (it hoped) would go to market for xenografts, and Akira Onishi of Japan's National Institute of Animal Industry cloned Xena from another pig fetus.

Both researchers aimed at producing a pig with organs that would not be rejected when transplanted into humans. Both will use batches of identical, cloned pigs in experiments to suppress a sugar called alpha-1,3-galactose, which resides on pig cells and causes humans to violently reject transplanted pig organs. "Getting rid of this sugar may go a long way towards the ultimate goal of providing an unlimited supply of compatible pig organs for human transplantation," said a scientist at Roslin.[2]

Scientists in Taiwan in 2002 cloned two identical descendants of an alpine goat that was a champion milk producer. Japan had already imported a herd of alpine goats that adapted well to its hotter climate. A year after their birth, visitors met the healthy twins at an international biotechnology convention hosted by Taiwan.

In 2003 French scientists successfully cloned a rat. Cloning a rat was a boon to scientific research because rats are used extensively in research, for example, testing drugs for hypertension and diabetes. Cloned rats will also serve as identical sources for experiments in knock-out genes, where a gene is deliberately deleted to study what happens in the adult animal.

When Steve Stice (now at the University of Georgia) created the first cloned calves at Advanced Cell Technology, researchers used a technique inadvertently discovered by Neal Furst of the University of Wisconsin. Furst starved the eggs before attempting nuclear implantation. For rats, researchers learned to slowly mature the rapidly developing eggs before they implanted a nucleus.

CLONING CHAMPION DAIRY COWS

Scientists first produced a cloned Holstein cow on June 11, 1999, at the University of Connecticut at Storrs. The cow, named Amy, had a genotype taken from the ear of a prize cow named Aspen.[3] This asexual technique was used later by Japanese scientists to clone a prized bull. It bypasses the cumbersome, normal process of inducing superovulation, harvesting eggs, and introducing sperm.

Another extraordinary cow, Zita, produced ninety-five gallons a day of high-quality milk at two years of age. In 1997 Zita was ranked the number 1 Holstein in America.[4] Zita became the ancestor of many cloned dairy cows. Demand for Zita's genetic material was so great that in 1997, when Zita's owner visited Japan, the dairy industry there saw cloning as the best way to revive its ailing cattle industry.[5] The Japanese bought five of Zita's sons for $125,000.

As Zita aged, her owner worried about whether her cells would be potent in producing more embryos, and when he met a representative of ACT, which had created a new unit called Cyagra to cloned valued livestock, he agree to use Zita's cells to produce the first cloned calves. In early 2000, Cyagra used cells from Zita's ear to create twenty embryos. Winnowed to the best ten, these embryos were inserted into ten surrogate cows, where they gestated for thirty-five days, creating eight pregnancies. After 120 days, six died shortly before birth. Two healthy fetuses made it to birth, Cyagra Z and Genesis Z, on February 6, 2000. Two weeks before, Japanese scientists had announced the first grand-clone of a cloned bull, giving them three generations of the same genotype.[6]

A year later, Cyagra cloned a descendant from a prized Wisconsin cow named Mandy and then auctioned it off for $82,000.[7] Infigen, of DeForest, Wisconsin, also cloned a prized Holstein cow in 2001 and sold her for $82,000.[8] In 2002 Fidel Castro encouraged Cuban scientists to clone Ubre Blanca (White Udder), which had once produced a world record of 241 pounds of milk in one day (about four times the normal production).[9] Such clones might alleviate a shortage of milk in Cuba.

CLONED RACING MULES AND HORSES

The world took notice in May 2003 when a five-year project at the University of Idaho had a mare give birth to a normal mule created from a cloned embryo. Idaho Gem was the first member of the equestrian family to be cloned. Mules traditionally result from breeding a male donkey to a female horse. "A mule can't do it himself, so we thought we would give it a hand," said Gordon L. Woods, the lead scientist. Idaho Gem's birth occurred about the time that Americans embraced the horse Funny Cide, the gelding that won the Kentucky Derby and the Preakness but failed to win the Belmont and thus missed the Triple Crown. With equestrian cloning now possible, Funny Cide might have descendants.

Professor Woods rebred the parents of Taz, a famous champion racing mule, to produce a fetus that would have been a brother of Taz. In the pursuit of cloning, they aborted this fetus and inserted nuclei from its cells into 307 eggs, eventually producing Idaho Gem, the fetal brother of Taz. They had worked for over three years on this objective and experienced many failures.[10] Their breakthrough occurred when they began to focus on calcium levels during the transfer of the nucleus in the fluid surrounding the egg. About a month later, the same group of scientists produced a second twin, Utah Pioneer. Both foals originated from a fetal skin cell culture established in 1998 with Taz's mother and father. A third twin born in July was named Idaho Star.

Although not familiar to most people, mule racing finds favor in the American Northwest. Before Idaho Gem, Taz competed mostly against another champion, Black Ruby, which goes a long way toward explaining why owner Donald Jacklin invested $400,000 in the champion mule-cloning project. He plans to train all the cloned mules to race and soon a race will be run where the top six contestants will be genetic versions of Taz. Does that sound monotonous? So was Taz versus Black Ruby again.

Three months after Idaho Gem's birth, Italian researcher Cesare Galli achieved the birth of the world's first cloned horse.[11] Galli called the foal "Prometea" after the Greek god Prometheus. "Like Prometheus who took fire from God from Olympus, we hoped she would be brave to face these people who do not like what we do," he told reporters.[12] Galli may have a way to go before conventions change. The Jockey Club, which controls horse racing in America, now forbids the racing of any horse that was cloned, as does the American Quarter Horse Association.

Prometea came from a single skin cell of her ancestor. In a first, her genetic ancestor was also her gestating mother. Some skeptics had claimed that a female could not bear her own twin because some minor genetic difference was needed for successful gestation. This skepticism was proven to be superstition, not fact, as Prometea grew successfully for the full eleven months and arrived in perfect health. As usual, Galli had started with a higher number of pregnancies (four from seventeen implanted embryos), two of which spontaneously aborted and one of which miscarried halfway through.

Cloning is best thought of as a tool in the life sciences that, once perfected, can be used in many ways; in itself it is neither good nor bad. Although cloned horses currently are not allowed to race, horse cloning is a tool that allows for some new possibilities: geldings could be reproduced and their descendants could produce sperm. And although geldings can't reproduce and cloned horses can't race, what about a cloned horse from a gelding put out to stud? Would the sexually created descendants also be banned from racing or would they be allowed to race? The Jockey Club will soon need to answer such questions.

Horse cloning could also aid varieties of wild horses that are endangered, helping nature by reintroducing strong wild horses. Finally, many horse competitions do not ban cloned horses, such as jumping, dressage, harness and barrel racing. Perhaps the greatest demand for cloned horses might come from ordinary people who want a piece of greatness. If Seabiscuit were still alive, wouldn't demand be high for his cloned twin-descendants?

Of course, the nature versus nurture debate will rear its head here, and this might be an easy place to start real studies. People will swear right away that no clone of Seabiscuit will ever run as fast as Seabiscuit because of Seabiscuit's unique upbringing, experiences, and bonds with his jockey. Ditto for Secretariat, Alydar, and Seattle Slew.

But we shall see. People have a lot of emotion invested in their views. Since breeders know that clones of these horses have the genetic potential to be great, they might take extraordinary care to create exactly the right conditions for maximal potential. And they might get lucky very early or we might figure out the science very early, and regularly produce horses whose times match or beat their ancestors. If not, it will be an early test of how important the environment is to ultimate physical performance. And that too will teach us something important.

CLONED LIVESTOCK AND ABNORMALITIES

Critics of cloning such as MIT's Rudolph Jaenisch claim that because of reprogramming errors in the creation of the cloned embryo, no cloned mammals are normal, including Dolly and the champion cows described above. Such claims about Dolly have a political context and pedigree, which is worth understanding, so we digress for a moment to discuss the claim that Dolly was "six years old at birth" as well as the claim that "she died early because of cloning."

In addition to false claims that Dolly had not really been cloned from a differentiated cell, there was speculation, beginning when Dolly was three years old, that cloning caused her to age prematurely.[13] A study released by the Roslin Institute in June 1999 showed that Dolly's telomeres were 20 percent shorter than normal lambs at the same age.[14]

Telomeres are located at the ends of chromosomes and are speculated to be indicators of cell aging. They protect chromosomes from injury when cells divide. In most cells, the telomeres get shorter each time that cell divides. In humans, telomeres divide about twenty times over a lifetime, and in cows, sixty times. The number of times cells divide in a species is called the Hayflick limit. Telomeres are like a shell around an egg, and when the shell has eroded, this chromosomal egg breaks. This breakage may start many diseases of old age.

News circulated widely in 1999 that Dolly had shorter telomeres. "Three-year-old Dolly has 9-year-old body!" headlines screamed. But fewer headlines ensued when other researchers found that the telomeres of cloned calves were longer than normal, suggesting not premature aging but a fountain of youth. In August 2000, Dr. Robert Lanza, chief scientist at the famous for-profit firm, Advanced Cell Technologies (ACT) in Worchester, Massachusetts, announced that he had studied the telomeres of his cloned calves and found not shorter but longer telomeres. The team of famed Hawaiian researcher Ryuzo Yanagimachi also found the same good effects over five generations of cloned mice.[15]

Although Lanza's and Yanagimachi's results excited some attention in the press, they were generally lost in the mass of many announcements about cloned animals and stem cells. Unfortunately most people still have the false impression that Dolly was old at birth. A similarly politicized story made the rounds at Dolly's death. Lambs living free in pastures may live twelve or thirteen years, so when Dolly had to be euthanized at age six, critics predictably sharpened their knives and had a feast. Alan Colman, a scientist who worked at Roslin with Ian Wilmut and fled to Singapore for less-restrictive working conditions, said that Dolly's premature death was proof of the many dangers of cloning, "I think it highlights more than ever the foolishness of those who want to legalize (human) reproductive cloning. In the case of humans, it would be scandalous to go ahead, given our knowledge of the long-term effects of cloning."[16]

The British antiabortion group Life claimed that Dolly's early death showed why all research into animal or human cloning should be stopped as unnatural or perverse.[17] Even though the National Academy of Sciences concluded that cloned livestocks (and their milk) do not differ in any way from traditional livestock and senior FDA scientists came to the same conclusion, the director of food policy at the Consumer Federation of America, Carol Tucker Foreman, objected to introducing food or milk from cloned livestock into the American food supply: "When you say animal cloning, many people react as if you are at least opening the door to human cloning."[18] These are but two of several examples of anticloning groups who claim that human reproductive cloning is linked to human embryonic cloning, which is in turn linked to nonhuman mammalian animal cloning. Because all of them are linked conceptually and empirically to slippery slopes, none of them should be done.

As I shall argue in chapter 8, this is a major reason to look hard at the claim that human reproductive cloning is evil. If human cloning can never be good for humans, even if safe, then it brings down anything linked with it, such as embryonic or animal cloning. On the other hand, if such cloning is only bad because it's unsafe to try at the moment, then we can go forward with other kinds of cloning. Indeed, we should go forward, for that may be the only way to make human reproductive cloning safe.

Dolly's early death was most likely iatrogenic—caused by those around her, not a pathology in the way she was originated. In the philosophy of science, the Hawthorne effect states that merely observing a subject may alter the behavior of the subject from what it would be if no one were around. Dolly's life was a continuous Hawthorne effect, from her heralded birth to her unexpected death. Always sought out by tourists and journalists, Dolly learned to stand on her hind legs and beg for food, which she usually got. So she grew fat and, as any pet owner knows, a heavy dog or cat putting weight on its back legs can experience injury late in life. (I once had a spoiled golden retriever who learned to stand on his hind legs and snatch food from the kitchen counter. This behavior, plus being overweight, exacerbated a congenital hip dysplasia and led to hip pain at age sixteen.)

After six years of hysteria over cloning in the mass media, veteran journalists learned to discount the sky-is-falling talk about cloning—in this case, hyperventilations about Dolly's premature death from cloning. *Washington Post* science writer Justin Gillis wrote that Dolly was "the friendly but spoiled sheep" who would run to meet photographers toting cameras, put her feet up on a fence, pose for pictures, and demand food for her trouble. Dolly ate so much from the hands of visitors that she got fat, even by sheep standards. She contracted arthritis at a young age for a sheep, though it was never clear whether this was from too much picture posing or from the circumstances of her birth.[19]

Science writer Gina Kolata of the *New York Times* writes,

> For Dolly, life was good. Her first 10 months, she shared a stall with two other sheep, but she would grab their food and she soon began growing fat. She would assert her authority by turning over her trough as soon as she finished eating and placing her forefeet on it, puffing out her chest and preening. Deciding that Dolly was not good at sharing, the Roslin staff finally put her in a stall by herself. While most sheep are shy around humans and huddle at the back of their pens when visitors arrive, Dolly loved attention. She rushed to the front of her stall when she saw people coming by, bleating loudly.[20]

Dolly most likely died at an early age because she contracted a virus from being raised indoors. Her arthritis probably came from being overweight and standing too long in positions that are bad for sheep.

What about most cloned livestock? If Dolly and other cloned sheep are normal, what about cloned cows and bulls? Are most cloned cows and bulls abnormal? Jaenisch claimed that all cloned mammals are abnormal. If that is true, there is something odd about what has been going on in the livestock industry. One would think that companies paying $82,000 for a clone of Zita or Mandy would know by now that the dairy cow they bought was a freak and an abnormal producer of milk. But that is not the case. At least with cows, bulls, and lambs, reprogramming errors are not occurring.

Or perhaps the genetic flaw is subtler. William Rideout and Rudolph Jaenisch, coauthors of an article in *Science*, describe how apparently healthy cloned mice have malfunctioning genes:

> We have taken the viewpoint that what appears normal is not necessarily so. I think the same could be said with these cloned calves. Maybe a gene or two

> or a handful are not quite controlled in a truly normal fashion but that's likely to have little or no impact on what it is you eat or whether or not they have healthy offspring.[21]

The above quote shows that cloning *does not necessarily create* abnormal progeny. It holds promise that, as knowledge increases and as techniques are perfected, more and more normal progeny can be created among cats, dogs, pigs, and, yes, primates, giving us good safety data before human cloning is attempted.

Fifty years of stories about zombie clones, slave clones, mindless soldier clones, spare organ clones, and other stories assuming cloned humans are abnormal or subhuman have combined with the evidence of the first years of cloning that born human babies likely will have some congenital defects. Thcsc two fronts, weird clones from science fiction and initial abnormal babies from the first mammals cloned, together convince virtually everyone, even scientists such as Wilmut and Jaenisch, that human reproductive cloning is inherently wrong.

Mark Westhusin, a researcher at Texas A&M University, cloned eighteen identical Brahmin bulls when he worked at Granada, including the famous rodeo bull Chance, and the famous cat CC ("Carbon Copy") at Texas A&M. Westhusin thinks that Jaenisch's claim that cloned animals are abnormal due to errors of epigenetic reprogramming is false.[22] Westhusin's views are shared by animal researchers Steve Stice at the University of Georgia, who cloned the calves Charlie and George, and Jose Cibelli of Michigan State, who briefly grew a human nucleus in a cow egg while employed at ACT.

EATING FOOD FROM CLONED LIVESTOCK

One practical aspect of the controversy over abnormalities in cloned animals concerns eating the meat, eggs, and milk of cloned livestock. Despite upbeat reports in trade and financial periodicals, milk and beef from animals originated by cloning by the end of 2003 were not approved in North America's food supply.

Milk from cloned dairy cows may have already been inadvertently tested (apparently with no allergic reactions or bad results) because Granada Corporation and American Breeders Service made and sold hundreds of cattle clones and "milk from [such] clones was sold with no public outcry," says

Texas A&M veterinary professor Mark Westhusin. "There may still be some cows out there, producing milk."[23] A report in 2002 by the National Academy of Sciences (NAS) found problems with genetic enhancement of fish raised for food because of possible environmental harm to wild species, but it found no problem at all with allowing products from farm animals into the food supply because cloning did not alter the animal's basic genetic structure. "I think our message was loud and clear," said a biologist who was an NAS panel member. Virginia Polytechnic Institute biologist Eric Hallerman said, "The concern about food safety, we thought, was just way overblown."[24] Nevertheless, because of worries emanating from MIT, the Food and Drug Administration (FDA) has been reluctant to approve milk and beef from cloned animals, putting a hold on companies such as Cyagra of ACT in Massachusetts and Infigen of Wisconsin.

As in the battle over genetically modified veggies, critics of introducing milk and beef from cloned animals into the U.S. food market view concerns about safety as symbolic of much larger philosophical issues.[25] Foremost among these is the use of biotechnology in factory farming. Critics such as Jeremy Rifkin and Vandana Shiva see cloning as just another tool that capitalist agribusiness will use to commodify food production on a global scale, driving out small farmers, organic food, and community food markets.

At the end of 2003, we witnessed the powerful emotional kick of the dreaded C word as the FDA bowed to the public outcry from the critics and postponed allowing dairy products and beef from cloned animals into the food supply. A public campaign from the organic food industry urged, in regulating such food, use of the precautionary principle, a principle used in Europe to ban import of genetically modified vegetables until they were proven safe beyond reasonable doubt, an impossible standard to prove and a mask for trade protectionism.[26]

A month later, hundreds of people became very ill and a few died from hepatitis A contracted at a Mexican restaurant outside Pittsburgh. The culprit? Not genetically modified veggies or lack of hand washing by workers, but fresh scallions.[27] Turns out that scallions are washed very little before being cut up raw for salads and seasonings, and if they get contaminated along the way, it's a free ride for the pathogens right into the victim's gut. This raises questions about the safety of eating organic scallions, which are grown in manure-fed soil. Meanwhile, "USDA-inspected" meat at supermarkets continues to mean virtually nothing, as infected beef routinely gets made into hamburger.[28]

According to MIT's Rudolf Jaenisch, there is "no such thing as a healthy clone." Jaenisch's claim influenced medical-science writer Rick Weiss of the *Washington Post,* who wrote that it's one thing to eliminate livestock with abnormal tissue from the food supply, but how can food from normal-looking cloned animals be detected who have bizarre "patterns of gene expression inside their cells. As a result, some cells may be making different proteins than normal, or they're making proteins in different quantities than normal, or they're making proteins in different organs than they'd normally made." [29] Weiss implied such food should be off the table.

Despite Rick Weiss's fears about aberrant proteins, it is a principle of critical reasoning that it is impossible to rationally assess risk without making a comparison. Yes, eating food from the descendants of cloned dairy cows or cloned bulls might carry a small, unknown risk, but nothing in life is completely free of risk. Someone walking down the street could theoretically be killed by a meteorite. Asking someone to prove that something is risk free is not a rational request but an emotional appeal. Fear is evoked by stressing that risk indeed exists. But the rational question always is, How much and compared to what?

How safe would eating meat or milk be from cloned livestock compared to other risks involving food? Table 3.1 is instructive and is taken from David Ropeik and George Gray at the Harvard Center for Risk Analysis.[30] In it, "illness" is defined as requiring hospitalization.

As Ropeik and Gray write, "You are more likely to be affected by foodborne illness than almost any other risk in this book. Approximately 76 million Americans, roughly one in four, suffer food poisoning each year . . . foodborne illnesses kill approximately 5,000 Americans a year."

The authors also rate many other risks, which are instructive to compare. In table 3.2, they first give the likelihood of the average American's exposure to a harmful level of the substance (on a similar scale of 1 to 10,

Table 3.1. Risk of Illness/Death per Year to Americans from Ingesting Various Substances

Ingesting spoiled/infected meat (E. coli 0157:H7, *Toxoplasma gondi*, etc.)	116,000/500
Eating spoiled eggs (Salmonella)	15,600/550
Ingesting raw oysters, shellfish, sushi	20,000/124
Ingesting organic produce (E. coli 0157: H7, *Giardia lamblia*)	2,000/100
All foodborne pathogens causing illness or death (many more than above)	76 million/5,000

Table 3.2. Other Risks from Ingesting Substances on a Scale of 1 to 10, with 10 = Highest Risk

Ingesting genetically modified vegetables	.01/1 = .01
Ingesting aspartame and saccharin	.01/.02 = .0002
Ingesting alcohol	8.5/7.5 = 63.5
Ingesting tobacco	5/9 =45
Ingesting mercury	1/.02 = .02
Ingesting pesticides	1/1 = 1
Ingesting DDT	1/.02 = .02

with 10 being most likely) and a second figure for severity of consequences plus-number of victims. So the first figure gives the likelihood of bad effect and the second, if a bad effect occurs, the magnitude of the harm resulting.

Obviously tobacco and alcohol are by far the most harmful things people ingest: "One out of two of today's 47 million smokers who don't quit will die from tobacco exposure."[31] Similarly, "millions of people suffer [bad] consequences from alcohol, a huge number of which are the most severe consequence of all: death."

Given this comparison, I think that, even with existing uncertainty about eating dairy products and beef from descendants of cloned animals, the risks involved at most would be the same as those of eating genetically modified vegetables, which have no risk of death or disease and only a miniscule risk of allergic reactions. As such, risks of eating food from cloned animals would be the following: Ingesting beef from cloned animals .01/1 = .01.

Note that eating tuna or other fish every day may result in exposure to mercury, which is why pregnant women must be careful about how much they consume. Eating raw beet, alfalfa, or clover sprouts or organic apples on slow-food farms, where cows graze underneath apple trees and leave their manure on fallen apples, can also be hazardous because of deadly E. coli 0157: H7.

Overall, each day 100 million Americans eat hamburgers, sushi, and raw vegetables, and drink alcohol. All these risks are far higher than anything conceivable from genetically modified vegetables or food from cloned animals. Astonishingly, because of intense media coverage, many Americans regard food changed by biotechnology as more dangerous than traditional foods, such as beer and hamburger. A rational look at the evidence shows that is not the case. Genetically engineered growth hormone for cattle was much safer than cadaver-derived hormone, yet people fear biotechnology in the food industry. As David Ropeik argues, such coverage by the media creates perceptions of risk, which differ greatly from real risks.[32]

Here we see a disturbing new trend: the alliance of environmentalists and food naturalists against biotechnology. Fueled by an antirational romanticism and a yearning for the natural comfort foods of the hearth, the alternative food industry has quadrupled its sales and profits in the past few years as a result of its alarmist campaigns against GMOs (genetically modified organisms) and biotechnology. What is alarming is the constant appeal to emotion and use of sensationalism designed to influence the media in the membership of these groups, which includes many otherwise good-thinking professionals and educators.

Two days after the Organic Trade Association blasted the FDA on its preapproval of cloned food, the FDA scientific panel took another look at its own findings and reversed its previous go-ahead for cloned food.[33] As a result, it will likely be years before such food is approved for distribution, a major setback to investors and the advancement of biotechnology in agribusiness.

CONCLUSION

As already noted, cloned animals are not a sideshow to the main event. In many ways, they are the main event, and for good reason. If there is truly "no such thing as a normal clone," then we are a long way from safely cloning humans. On the other hand, if expectations and ideology influence how people see cloned animals, then time may tell a different story. The story of cloning animals, however, is not over, and the following chapter discusses other controversies in cloning animals such as cloning pets and endangered species.

4

Cloning Endangered Species and Pets

We had Smokey neutered at a young age. We wanted some of his offspring, so when we heard about cloning, we thought it was the perfect thing to do.

—*Mary Ann Daniel, www.savingandclone.com*

CLONING ENDANGERED SPECIES

In April 2003, speculation became reality when scientists in Iowa created the first healthy offspring through cloning a mammal of an endangered species—a banteng, a cowlike creature native to Java and Indonesia.[1] Worldwide, the population of bantengs has dwindled 80 percent. Owing to the destruction of natural habitats and hunters, large herds no longer roam through forests and fields in Southeast Asia. One day soon, this kind of cloning will also be used for Tasmanian and Sumatran tigers, African bongo antelopes, the giant panda, and maybe the Dodo bird.

Remarkably, the baby banteng came from cells taken from the ear of a captive banteng in the San Diego Zoo that had been frozen for over two decades. The ancestor had died in 1980 before it could breed, which happens rarely in captivity. The gestation and birth occurred in Sioux City, Iowa, at Trans Ova Genetics. The company also produces drugs and proteins from genetically enhanced cows.

In 2001, a dairy cow in Iowa named Bessie was surrogate to an embryo of a gaur, another endangered species. As with bantengs, gaurs don't often mate in captivity. Although the baby gaur appeared healthy at birth and stood on its legs and explored its surroundings, it soon contracted dysentery and died.

Like other kinds of mammalian cloning, the cloning of the baby banteng was inefficient: scientists affiliated with Advanced Cell Technology (ACT) started by transferring dozens of banteng embryos into beef cows but brought only two calves to birth. The second banteng calf, twice the normal size at birth and unhealthy, was euthanized. The firstborn banteng has done well and will grow to its standard weight, nearly 1,800 pounds.

For twenty-five years scientists at the San Diego Zoo have pushed Frozen Zoo, which stores frozen tissue from the world's rarest animals, including condors, pandas, and California gray whales.[2] Coupled with the San Diego Wildlife Preserve north of the zoo, the San Diego Zoo has a good intermediary environment for raising and monitoring reintroduced exotic species, making sure they are viable to before releasing them into the wild.

Betsy Dresser and Earl Pope work at the Audubon Nature Institute's Center for Research of Endangered Species, which is trying to clone bongo antelopes, gorillas, and exotic cats. The center lies on 1,200 acres on the west bank of New Orleans. Its strength lies in assisted reproduction with frozen embryos and interspecies embryo transfers.[3] At the end of 2003, it had cloned several kittens of an endangered species, the African wildcat.[4] The first, Ditteaux, was born August 6 and two more followed on November 15. All three kittens were ancestors of Jazz, a thawed embryo gestated by a domestic cat.

In July 2003 scientists in a Chinese biotechnology company, in collaboration with the Institute of Zoology, Chinese Academy of Sciences, cloned twin cows by somatic cell nuclear transfer (SCNT).[5] Scientists there are also using rabbits to gestate panda embryos.

Impartial observers might have predicted that environmentalists would be elated at this new tool to preserve endangered species, but most greeted the news with disdain. Why? Because they thought that cloning endangered species illustrated a trend toward reductionist science. Cloning the banteng represented a trend of seeing it as nothing but its genes, not as part of an integrated environment. The World Wildlife Fund said the only correct way to preserve bantengs was to preserve natural habitat. Cloning moved in the opposite direction because it could excuse destruction of natural habitat ("if they become extinct, we can always clone some later").

As mentioned in chapter 3, this response illustrates a disturbing new alliance against science and biotechnology by environmentalists, animal rights groups, and bio-Luddites. None of these groups sees biotechnology as providing useful tools to manage the environment, probably because they do not trust business or government to do so. But that is like arguing we shouldn't accept electricity because it led to the electric chair or shouldn't study physics because it led to nuclear bombs. As I will stress throughout this book, a tool is just a tool, and no tool should be discarded because some people fear its possible uses.

The response of environmentalists to the cloning of the banteng calf also illustrates the *purity problem* in ethics: how pure should we be in our expectations of our fellows? In the previous section, we saw a similar problem about raising animals for food. Should we hold out for universal adoption of the highest standards, avoiding all compromises? Or should we dirty our hands to compromise with the real world, especially if that means resigning ourselves to impure results?

It's hard to imagine anyone who would not prefer saving endangered species by preserving their habitats. If the question is simply, Should we save endangered species by preserving their habitats? everyone will answer yes. But in the real world, people value more than one thing. Axiological monism, asserting that humans should only value one thing in life, is not a tenable assumption, be it about trees, embryos, or drugs.

If we then ask, Should we save endangered species by preserving their habitats, even if it costs a million Brazilians their jobs? Or even more pointedly, Should we save endangered species by preserving their habitats, even if it costs the lives of a million Africans? Then the answer is far from obvious. This question is not as crazy as it sounds. Greenpeace opposed introduction of genetically modified crops to preserve the environment in Zimbabwe, even though its people were starving.

When confronted with such trade-offs, people characteristically deny them; for example, Greenpeace denies that GM (genetically modified) veggies will help feed the starving and thinks that starvation can be cured in environmentally friendly ways. But the question must then be pushed: Yes, but *what if*, the only way to save a million (jobs, lives) is to sacrifice the environment? Is it always right to save the environment?

"As science, cloning is fascinating, but as conservation, it's a complete farce," says Karen Baragona of the World Wildlife Fund. "Cloning one banteng does nothing for its habitat or the wildlife that shares it."[6] Well, yes and no. Yes, cloning does nothing for habitat conservation or existing wildlife, but of course, that is not the point. It's just a tool that might help.

Consider the California condor, almost extinct in the Grand Canyon but reintroduced. When almost extinct condors did not mate and reproduce in the canyon, biologists created embryos/eggs and hatched them in laboratories and then introduced the young birds to the canyon. Now a few dozen are almost at the point of breeding naturally. Wasn't that a good use of scientific labs? Isn't cloning such condors just one step away from what we did with the condor embryos? Why would it be bad to use this new tool?

Preventing the destruction of the environment is not an absolute value, even in sophisticated North America; it is just one of many important values. Developing countries may give almost no value to preservation of natural jungles, meadows, and habitats of endangered species compared to how they value creating a healthy economy. For that very reason, having a new tool to preserve endangered species allows us later to help such countries restore species they lost.

Saving an endangered species will soon force the questions of which species to preserve and, more importantly, why we should preserve endangered species? In the history of evolution, species become extinct all the time in the competition for food and resources. In one sense, *that is just what evolution is.* When people want to preserve every species that exists today, even when it is endangered, they want to thwart evolution. Is that good?

The most understandable view says species shouldn't unintentionally become extinct because of man's careless destruction of habitat. In such cases, if habitat cannot be reconstructed, cloning animals might give them a chance to live again. But if the forces that brought about the original demise are not changed, cloning will only be a stopgap measure. In medicine, halfway measures often don't prevent death. Sometimes, however, they buy time and allow the body to heal. Can't new tools sometimes help the environment to heal?

Finally, extinct species with no live cells can't be re-created by cloning (the Jurassic Park problem). Several years ago, the Discovery Channel collaborated with an entrepreneur to try to clone a baby woolly mammoth from one frozen in Siberia. But we know from the fledgling science of cryonics and cryopreservation that freezing biological cells destroys them by shattering them. So no live cells exist to clone a woolly mammoth from; nor is it likely that any human will ever be brought back to life using current techniques in cryonics.

THE ETHICS OF CLONING PETS

As we move from the cloning of animals that we eat for food ('livestock") to cloning endangered species, and from there to cloning genetic replicas of our pets, we move across an emotional continuum from bare interest to personal involvement.

In 2000, Arizona billionaire John Sperling gave $2.3 million to researcher Mark Westhusin and three other researchers at Texas A&M University, in a joint venture with Bio-Arts and Research Corporation (BARC) of California, to clone the genotype of his dog, Missy, a mix of border collie and husky.[7] The Missyplicity project has been going on since 1997 but has failed to produce viable canine offspring because dogs are much more difficult to work with than sheep or cattle.

Because of an inoperable esophageal tumor, Missy was euthanized in July 2002. But instead of giving up, Sperling gave the project $3.7 million more. That grant led directly to creation in December 2001 at Texas A&M of the cat CC (Carbon Copy), who to date has been perfectly normal.

Controversy broke out over the fact that CC's ancestor, Rainbow, was an orange calico, but CC had a striped gray coat over a white base. Wayne Pacelle of the Humane Society harrumphed that the difference showed that cloning didn't work.[8] CC's originator, Mark Westhusin, replied that he knew beforehand that CC's coat color would be different because coat coloring and pattern is controlled by random factors in genetic reprogramming.

At the end of 2003, Westhusin announced creation of Dewey, the world's first cloned white-tailed deer.[9] Dewey was named after veteran Texas A&M researcher Duane Kraemer. Westhusin and Kraemer hoped that success with Dewey would help them clone other wild animals, including endangered ones.

All kinds of people are encouraged by these projects and have sent researchers samples of pet DNA. BARC and its president, Lou Hawthorne, started Genetic Savings & Clone, Inc., to store samples of dogs toward the day when dog cloning is reliable and safe. As of 2002, two hundred samples were stored at a cost of about a thousand dollars each.

Cloning cats will likely lead to allergen-free cats, enabling more people to have cats in their homes. Professor Xiangzhong Yang of the University of Connecticut is working with the company Transgenic Pets of Syracuse, New York, to produce such a cat.[10] An estimated 30 million people in North America are allergic to cats. Most react to an allergic protein secreted in a gland under the cat's skin that is covered by fur. Researchers will eliminate the cause of the allergy at its source by producing a cat with a knock-out gene that differs in no other way from traditional cats. The company plans to sell them for a thousand dollars each.

Benefits for humans include being able to own a cat and not needing to take medicine to reduce the effect of the allergies. Benefits for cats include 30 million new potential owners and a better relationship with them. Skeptics and radical environmentalists, of course, decried the venture, even though domestic cats and dogs have been consciously bred for centuries for specific human desires.

Controversies about cloning pets illustrate differences between the spheres of law, public policy, morality, and personal life. Where humans unequivocally harm another, laws exist to prevent harm (laws against running red lights). Where we want to influence human behavior, we make public policy (we give tax credits for adopting disabled children). Morality also concerns actions that might harm or benefit others, such as feeling wronged by others smoking next to us in public places. Finally, personal life concerns actions by me, or by my family, that affect others very little, if at all (smoking in my bedroom).

Deciding to clone the genes of my deceased pet falls in the sphere of personal life. As such, the decision should fall outside criticisms from morality or public policy. Nevertheless, criticisms from such external areas are always raised, for example, that cloning pets deprives animals in shelters. (Although such adoptions may not do much good, as they encourage people to abandon their pets or allow them to breed in the false belief that most pets at shelters get adopted. So even criticism from public policy has its perils.)

Is it people's personal business how they spend their money on pets? Whether they want to try cloning to re-create a loved pet's genotype? Not so, says the American Humane Society's Wayne Pacelle, who called the Missyplicity Project "perfectly awful" when so many dogs get euthanized while waiting to be adopted.[11] Representing People for the Ethical Treatment of

Animals (PETA), Michael W. Fox, the well-known animal rights advocate and veterinarian, says, "Dog cloning is a sentimental self-indulgence for those who can afford it."[12]

Fox also objects to the sixty-one female dogs that supply the eggs into which nuclei of Missy are injected to create embryonic clones of Missy. "Those dogs in the Missyplicity dog colony are hormonally manipulated to ovulate faster than normal. Eventually, that's going to wear them down to the point that they'll develop diseases earlier and start dying sooner. We've only recently gotten down from the trees, and we're already playing God. We need to think about our responsibilities to animals."

The question of cloning pets resonates with us because it is emphatically not the same as cloning livestock or animals for the dinner table. Cloning my pet dog or cat enters the realm of my personal life, my home and deepest feelings, separate from the public realm of law, public policy, and maybe morality. At the same time, if cloning of pets is possible, it desensitizes us to (what Leon Kass called) the "repugnance" of cloning. After all, if cloning the genotype of my beloved cat of eighteen years is permissible, why isn't it also permissible to clone the genotype of my suddenly deceased eighteen-year-old child?

Cloning animals obviously raises ethical issues, whether we're cloning deceased pets, livestock for milk and meat, or members of endangered species. But what happens if we go one step beyond and attempt a border crossing of two species? Are animal cyborgs morally acceptable? Would it be environmentally permissible to introduce cross-species into our world? Chapter 15 explores these questions.

REFLECTIONS ON CLONING ANIMALS

A final reflection for these two chapters on animal cloning. Many ethicists and critics now seek a consensus in the United Nations to criminalize embryonic and reproductive human cloning. Even if one agrees with such a ban on reproductive cloning, does work on animal cloning show a ban on human cloning to be a good thing? Suppose that Dolly's idea man, Steed Willadsen, or her breeder, Ian Wilmut, had sought a worldwide consensus before proceeding to try to create Dolly.

Would they have had any chance of getting such a consensus? Wouldn't skeptics have told them to try other kinds of science, since we "knew" then that it was a law of nature that mammals cannot be successfully born by the asexual methods of cloning? Would Robert Edwards and Patrick Steptoe have gotten a go-ahead before they achieved Louise Brown's birth as the first baby originated by in vitro fertilization, especially after it was learned that many of their embryos aborted?

Fortunately, few people think it necessary to get consensus on cloning different species of animals before going ahead, although after each species is successfully cloned, critics are legion. But, as we shall see in a Chapter 15, almost everyone feels differently about creating by cloning a mixed-species transgenic animal, such as a "geep" (mixture of sheep and goat) or a pig that may be more intelligent than a human adult with an IQ of 50.

5

Psst! Raël Wants to Sell You the Brooklyn Bridge

Paracelsus, a physician, pretended, not only to make gold and to confer immortality, but to cure all diseases. He was the first who . . . attributed occult and miraculous powers to the magnet. . . . His fame spread far and near.

—*Charles McKay,* Extraordinary Popular Delusions and the Madness of Crowds

By now, almost anyone who has heard of cloning has heard of Raël and his followers, the Raelians, who are irretrievably associated with human cloning. So a topic that had already suffered from fifty years of scary science fiction now suffers from association with an exotic cult leader who claims that humans were created by aliens whose religion is science. (Well, sort of pick-and-choose science, since Raël rejects evolution.)

Surprisingly, some people believe the Raelians cloned a human baby. One of their reasons is that the media wouldn't devote so much attention to the Raelians if their claim were a hoax. After all, the *San Francisco Chronicle* put their claim on the front page, just as it did when news broke that Ian Wilmut had cloned Dolly. Nevertheless, the Raelians are unequivocally and unquestionably a hoax.

To understand the depth of the hoax, let's review the history of Raël, his henchwoman Brigitte Boisselier, and their organizations, the Raelians and Clonaid. Both truth and ethics emerge from the details, especially the cumulative weight of evidence supplied by many details in context.

After the announcement of Dolly's cloning in 1997, Raël set up in the Bahamas a private company, Clonaid. Little more than a post office box, Clonaid cost very little to incorporate. Doing so allowed Raël to make money off the gullible by offering to clone a child for $200,000.[1] Several different times, Raël and Boisselier have admitted what they're doing. His arrogance is his downfall: it makes him want to brag about conning the media. It's as if he needs to step out of costume once in a while and say, "Look, Ma! Can you believe I'm getting away with this?" In his 2002 book, *Yes to Human Cloning*, he wrote, "For a minimal investment of $3,000 in U.S. funds, it [Clonaid] got us media coverage worth more than $15 million. . . . I am still laughing. Even if the project had stopped there, it would have been a success."[2]

For Raël, success is having a good time, being on television, and making fun of people's gullibility. Raël has never denied that he might be a huckster. When invited to testify before a committee of the House of Representatives in 2001 (before which I also testified), Raël was invited to maximize coverage by the national media (see preface). As University of Pennsylvania bioethicist Arthur Caplan testified that day, "The House invited Raël down to show that cloning was kooky and dangerous and imminent."[3] Caplan added, "And Raël allowed himself to become the poster boy of the anti-cloning movement in order to gain publicity."

Actually, members of Congress allowed Raël to make himself and Raelians the poster children for cloning. Raël knew exactly what he was doing, as did Boisselier. Raël shamelessly agrees to all this: "But we are winners in either case, even if cloning is banned. We are winners because our organization had worldwide media coverage."

When interviewed in early 2003 on *Dateline,* Raël admitted that it didn't matter to him if his sidekick, Brigitte Boisselier, had lied about the birth of the first human child from cloning: "It's a win-win situation for us. If what Dr. Boisselier is saying is true (that she has originated a human child by cloning), we are winning. If it's not true, we are winning anyway."[4] Why does Raël think he is winning even if the Raelians have not created a baby by cloning? Because, Raël gleefully says, "professionals estimate we received $500 million worth of free media publicity."

Raël dropped his mask long enough to reveal his real motives: marketing his cult, selling his alleged services to gullible buyers, taking hefty deposits for phony services, enjoying international fame, and having dozens of reporters hang on his every word. When the interviewer on *Dateline* noted this, Raël became indignant, objecting that he was not being treated like a priest or rabbi, implying the interviewer was a religious bigot.

Spammers make money sending millions of junk e-mails advertising products such as herbal Viagra and pornography. If only .001 percent of recipients respond, they make lots of money because the cost of sending the e-mail is trivial.

In late 2003, two reporters for *Journal de Montréal* infiltrated the Raelians, who are based south of Montreal near the Vermont border, claimed that Boisselier had mocked the credibility of the media in a private meeting with her supporters: "Come, my good journalist friends, ask me if we did all that to have the free publicity. Yes!" she told her fellow Raelians.[5] "When I played games with the journalists . . . everything turned into a circus." Longtime cloning activist Randolfe Wicker, who met Raël during a 1999 visit to his headquarters, likens his hoax to the one Orson Welles perpetuated with *War of the Worlds,* when radio listeners believed they were hearing an actual invasion of Earth by aliens.[6]

Raël has also been spamming women for a long time. In 1998, on his farm in Canada near the Vermont border, he announced that the Creators would soon return to earth and that pious women, especially beautiful young women, could join the new Order of Angels to help prepare their arrival by having sex with prophets of the Creators such as himself.[7] He had already had success with this scam in France, after giving up a career as a sports writer covering car racing. Raël started a movement that had female followers attending seminars in the nude as Raël addressed them and they all experimented with sensuality.[8]

As Raël and Brigitte spammed the globe with their cons, undoubtedly .001 percent of people responded. In the summer of 2000, a California lawyer, Mark Hunt, gave Clonaid between $200,000 (he said) and $500,000 (Boisselier said) to set up a secret human cloning lab to re-create the genes of his dead son.[9] Raël collaborated with Boisselier, holding a press conference where he introduced fifty female Angels willing to be surrogate mothers for the new cloned baby.

A television crew later discovered that Boisselier's lab was in an abandoned high school in the desolate Appalachian town of Nitro, West Virginia. White became disillusioned and demanded his money back. Boisselier rarely went to the abandoned high school, and no personnel ever worked there. After the school had shut down its ninth grade biology lab, no one had even bothered to even clean it. In short, no evidence whatsoever showed that any type if cloning was ever attempted in the lab, much less human embryonic cloning, much less implantations of cloned human embryos in willing human surrogates. Too late, Hunt realized he had been suckered: Clonaid was a pyramid scheme fueled by media exposure.

Raël and Boisselier never disguised their intentions. But many people had responded to their previous con and invested hundreds of thousands of dollars in their fake lab in West Virginia. Why shouldn't the same con work again?

Despite their history of fraud and deceit, when Boisselier announced that Clonaid had cloned a human baby between Christmas and New Year's in December 2002, all the major news organizations gave them nonstop free publicity for a week. Organizations such as CNN News, MSNBC, and the Associated Press couldn't resist a story involving aliens, cloning, and babies. Indeed, Boisselier and Raël received so much coverage during the last two weeks of 2002 that many people figured that where there's so much smoke, there must be a fire. Time and time again, when I explained to students and neighbors that it was a hoax, people would respond, "Then why would television news reporters spend so much time on it? Don't they investigate stories before they report on them?"

The sad answer to that question is no, especially under three conditions. First, it's the most sensationalistic story of the decade; second, Italian physician Severino Antinori was also promising to deliver a cloned baby by January 1, 2003 (he never did, nor has he ever produced any evidence that he had a late-term cloned fetus); third, the last week of the year is the deadest time of the year, when no normal person wants to be in the newsroom. As *Time* senior science writer Nancy Gibbs put it, "The science circus comes to town when a group like the Raelians claim to be cloning children, announcing one arrival just in time to fill the holiday news vacuum."[10]

Montreal professor Susan Palmer, who has followed Raël's career, is convinced that Raël has been "having the time of his life."[11] She says, "He's a playboy and a sportsman and a social satirist." As for his tendency to gather female groupies around him for sex, Palmer says that the only harm the women do "is to their marriages."

Good Morning America science reporter Michael Guillen also lent credibility to the Raelians. Either duped by the Raelians or hoping to make so much money he wouldn't need his day job, Guillen fronted for Clonaid and implied he had evidence that their story had credibility. Earlier Guillen had demonstrated a willingness to risk his credentials as a science reporter and hawked sensational claims that were backed by little real evidence but boosted television ratings.

Although Boisselier promised to provide DNA, she never did. Citing the convenient excuse provided by a Florida lawyer's self-promoting suit to gain custody of the alleged baby, Clonaid said the parents had refused DNA testing to avoid producing evidence for the suit.[12]

But the Raelians never produced a baby, never produced the surrogate mother, never produced the genetic ancestor, and, most important of all, never produced DNA of the ancestor and the baby. The last would have been the easiest way to prove their claim and a way compatible with protecting the privacy and rights of ancestor and baby.

Some ethicists should have been embarrassed for condemning what the Raelians had done, because in doing so they gave credence to the claim that the Raelians had actually tried to clone a baby. The appropriate response to a hoax is not to condemn it, but to make fun of it.

There had been earlier hoaxes about cloning: an Italian physician named Petrucci falsely claimed to have delivered an in vitro baby years before Louise Brown was born in 1978. An author named David M. Rorvik wrote a sensational book, *In His Image*, claiming to describe the secret cloning by an alleged scientist Darwin of the genotype of an alleged industrialist named Max, gestated by an alleged Sparrow.[13] Years later, a judge in Philadelphia judged the book to be a hoax and assessed damages to Lippincott and the author.

You would think that people would not buy books by such an author or, at the very least, publishers would not print further books by him. But Rorvik has published several more books about medicine, babies, and health (remember his name!).[14]

In 1998 Chicago physicist Dick Seed announced he would try to clone himself and gave many television interviews. The media and the public never seem to understand that a lot of people lead boring lives and like the sudden attention that claims about cloning humans always bring, such as free flights to New York to stay in luxury hotels. Why do you think people go on the *Jerry Springer Show*? In July 2002 Clonaid of South Korea claimed it had cloned a human embryo and implanted it in a voluntary human surrogate.[15] This group has even less reliability than North American Raelians.

Unfortunately, because most people did not understand that the Raelian claim was a hoax, anticloning politicians who must have known it was a hoax deceived them. Senator Sam Brownback (R-KS) called the Raelian claim a sign of "clear and present danger" that someone was about to try to clone a human unless the federal government made it a crime to try to do so.[16]

Cosponsor of the bill to outlaw all human cloning, Florida member of Congress and physician Dave Weldon used the Raelians to explain why all human cloning should be banned: "If you allow embryo cloning in research labs because of its supposed great potential, you're going to have all these labs with all these embryos and it will be that much easier for people like the Raelians to try to do reproductive cloning."

As I write this, more than a year has passed since the Raelians claimed to have produced a cloned baby and since Severino Antinori promised the same. Panayiotis Zavos has repeatedly promised couples that he can clone a child, but like Raël, has no real lab and is only interested in being before the cameras and asking for money for his talks on the lecture circuit or for taking deposits of $80,000 from gullible couples grieving over a deceased child.[17] Now we know they were all posturing, turning their fifteen minutes of fame into fifteen months. Let's be sure they don't get fifteen years!

In the end, Orville Schell, dean of journalism at the University of California at Berkeley, got it correct when he said, "The real story is the story about the story."[18] This chapter has been about that story and perhaps, because of it, fewer people in the future will pay attention to Raël and his minions.

6

Does Cloning Harm the Souls of Cloned Children?

I might be a twin, but I'm damned sure not a clone.

—*Englishwoman, 1998*

In both public and scholarly discussions of cloning humans, speculation abounds about what life would be like for a person originated by cloning. For example, how would such a person feel when he learned that his genes were the same as his genetic ancestor's? Some critics suggest that a cloned person would have an insecure identity. Other critics predict that parents might have difficulty raising a cloned child if, say, the mother felt attracted to a younger version of her husband. Others suggest that cloned people would feel, or would be made to feel, like manufactured objects, commodities, or, in Kantian terms, as mere means and not as ends in themselves.

Such psychological objections figure prominently in objections to reproductive cloning. Even if scientists learn to avoid abnormalities through better techniques, the wellspring of opposition to human cloning will remain, fueled by worries about the mind and feelings of the person cloned and the effects of his origins on his family.

Leon Kass, chair of the President's Council of Bioethics, focused the council report on just such aspects of human cloning. Fellow council member Francis Fukuyama similarly did not rally his objections to human cloning around the number of fetuses miscarrying or babies with abnormalities, but on claims about what cloning humans would do to the psychological essence of humanity.

MADE, NOT BEGOTTEN

In explaining why cloning might be bad psychologically for children, several commentators sum up one objection by saying that the child would be made, not begotten. The report of the Bioethics Council emphasizes this point:

> Procreation is not making but the outgrowth of doing. A man and woman give themselves in love to each other. Yet a child results, mysterious, independent, yet the fruit of the embrace. Even were the child wished for, and consciously so, he or she is the issue of their love, not the product of their wills; the man and woman in no way produce or choose a *particular* child, as they might buy a particular car.[1]

This description of the difference between sexual and asexual human reproduction, which is supposed to be the basis for explaining the immoral nature of cloning, is incorrect in two major ways: it is both too narrow and too broad. It is too narrow because procreation covers a much broader class of acts than those in which love is involved. Procreation can be just sex, and unfortunately babies do result from sex without love. Children result from the "embrace," but not in a mysterious way. Hence it is incorrect to say that a child issues from love because there may be only sexual passion involved.

On the other hand, the description is too narrow because it excludes the many couples who do not conceive easily but must plan and even work at conception. According to the latest statistics, a woman's ability to conceive with her own eggs drops dramatically around age twenty-seven, and so many couples who first try to conceive in their late twenties or early thirties have difficulty doing so. Indeed, about one couple in twelve cannot conceive after a year of trying.

More important, if willing, planning, or trying made the creation of children bad, then all children created by assisted reproduction would have been wrongly created. Similarly, children would be suspect if created when a woman took her temperature to judge ovulation and hence the best time for introduction of sperm.

Clearly the author of the report offers a romanticized view of sexual reproduction, necessarily filled with love, mystery, and lack of conscious planning. However laudable that view is, and indeed I find it a happy ideal, it is not one that can bear the weight of claiming on behalf of society that sexual reproduction is morally superior to asexual reproduction.

Cloned children are referred to often by critics as "made," "designed," or bought as "commodities." Do those analogies hold up? In reasoning by analogy, it is important that the two things being compared share as many properties as possible except for the one thing in which they differ. That is why it is a logical mistake to compare reproductive cloning to the production of cars on an assembly line (not to mention the intended invocation of *Brave New World*) or to buying and selling stocks on the New York Stock Exchange. It is a similar mistake to compare it to atrocities such as ethnic cleansing or weapons such as the atomic bomb. Reproductive cloning shares as much with these items as it does with basket weaving.

I don't think cloning is correctly compared to "manufacturing." I once taught in a building next to a Swingline Stapler factory in Queens, New York, and I can personally aver that human gestation of a cloned human embryo by a typical woman has nothing in common with a factory but almost everything in common with gestation of a sexually created embryo.

The frequency with which critics of cloning refer to it as "manufacture" attests to the power of the opening image of *Brave New World*, not the nature of cloning. This book is the mother lode from which many movies of biological or medical fiction derive. They scare us because babies emerge mechanically and uniformly, implying they are not wanted, cherished, nurtured, and bonded to their expectant parents. And yet, in contrast, the reality of reproductive cloning is that it may well be an excellent way to create children who are wanted, cherished, nurtured, and bonded to their expectant parents.

Perhaps reproductive cloning should be compared to gambling, which conservative religious people consider sinful and psychologists warn can express inappropriate behavior, especially where it becomes addictive. Nevertheless, we tolerate and regulate gambling, taxing it for schools and other public goods. Should we think of cloning like gambling? I don't think so. Even this comparison is way too dissimilar. Reproductive cloning shares nothing with gambling other than being condemned by some moralists at a particular time in history.

The best comparison is with a specific kind of in vitro fertilization where an embryo is created sexually from a man's sperm and woman's egg and then a primordial cell from that embryo is transferred to a younger egg for fusion, creating a slightly different embryo (with mitochondrial genes from the host egg). This technical comparison has the virtue of having only one thing different between the two processes: one transfer is from an embryo created sexually, the other transfer is from an embryo created asexually. The first process has been used in America and China to create babies for women whose eggs could not sustain gestation. The second process is reproductive cloning.

CONFUSED IDENTITY

A person of asexual origins would not know who he is, say Leon Kass and others, in the same way that is true for children created sexually. If a couple uses the genes of the man as a base for a male child, then the resulting child will be both twin and son to the father.[2]

Freud made us familiar with the concept of projection, in which we project our own secret wishes onto others and then denounce them in others, like the minister who constantly denounces lust in his congregation but secretly struggles with it himself. So for Kass, the real problem for the child cloned is the inability of the parents to love him as an ordinary child: "Virtually no parent is going to be able to treat a clone of himself or herself as one treats a child generated by the lottery of sex."[3]

Leon Kass, Ian Wilmut, and Arthur Caplan worry about incest between the child cloned and the opposite parent. "And what will happen when the adolescent clone of Mommy becomes the spitting image of the woman with whom Daddy once fell in love? In case of divorce, will Mommy still love the clone of Daddy, even though she can no longer stand the sight of Daddy himself?"[4]

It is sometimes said that people shouldn't write about very personal subjects because in doing so they reveal too much about themselves. Or perhaps these writers are bravely going where few writers care to go. In any case, if Freud is correct, it is likely that many men do have sexual feelings toward their daughters, who often do look like a young version of their wife. (What happens if, because of the sexual lottery, they look *better* than the wife at the same age? Is this a reason to ban sexual reproduction?)

As philosopher Bonnie Steinbock points out,

> We should ask whether it is the biological connection that prevents men from having sexual impulses towards their daughters, or the social role of father? If absence of genetic connection makes sexual feelings more likely,

> there would be an increase in father-daughter incest when a sperm donor was used or the child was adopted. . . . If the absence of genetic connection in these cases does not interfere with the appropriate father-daughter feelings, why should it in the special case of cloning?[5]

There are two kinds of writers on this subject. The first finds it unimaginable that the father or mother has sexual feelings toward a child of the opposite sex. The second thinks that such feelings do exist, but social roles and duty almost always restrain them, in addition to the lack of reciprocation from the child. Freud says that such feelings do exist between parent and child, but parents are socialized to repress them.

A similar claim of psychological harm concerns being a delayed twin of a genetic ancestor. At present, the significant others of twins who exist now shows us that twins are not interchangeable. Girlfriends do not treat boy twins as interchangeable. This is where Kass-the-emotivist gets lost. His emphasis on a visceral, emotional reaction to cloning leaves him with few rational responses. If ethics is merely how we feel and if men are attracted to their daughters, where does that leave Kass? His only solution is not to create the daughters in the first place.

Ultimately these writers beg the question against cloning by assuming that some kind of *special* attraction is created by cloning the mother's genes that would not be present in a girl created sexually. Fukuyama says a father will want to have sex with the young female clone of his wife and because she is a clone, he may actually do it. Fukuyama seems to be saying that every father wants to have sex with his daughter but refrains from doing so because of her normal human status. A clone, to Fukuyama's thinking, is not a normal human but a zombie or slave who can be sexually exploited. So for her own protection, we will not create her, just as we won't allow slaves to be created.

What is obviously wrong with this way of thinking, among other things, is the idea—so powerful in popular culture—that a child created by cloning would be subhuman or would have less than the full rights of a normal human child and could be used as a sexual plaything.

CLOSER SCRUTINY

Kass thinks that a child created by cloning will lead a life under close scrutiny as observers compare the life of the child to that of his ancestor.

This observation makes all kinds of assumptions, namely, that parents will reveal the child's unique origins and their expectations of him, even though it would not be in the child's best interests to do so.

Most objections of this kind are reductionistic in multiple senses and commit all the sins of reductionism. They assume that genes are everything, that parents will see genes this way, that the cloned child will see things this way, that his friends and relatives will do so, and that society will be reductionistic. According to such reasoning, parents would closely scrutinize the child of, say, Meg Ryan, expecting her to be like Meg Ryan's real-life phenotype. When the child grew up and heard she was from Meg Ryan's genotype, she would see herself as being totally determined by these genes. So would her friends, teachers, and extended family. Society itself would expect her to express the characteristics associated with those Meg projects in the roles she accepted as an actress.

The same reasoning explains why interracial marriage should be banned. The parents will see their child as the object of social prejudice and the child also will see himself this way, as will his friends and relatives, as will a society of racists. Ergo, ban interracial marriage. These arguments reify ignorance as reason, prejudice as justification. Because some people react badly, should such reactions rule public policy?

IS EVERYONE AN EXPERT IN CHILD PSYCHOLOGY?

Cloning exposes many aspects of the conventional wisdom, often in unflattering ways. It forces couples who would originate a child by cloning to give reasons publicly in a way not demanded of couples who sexually reproduce. Cloning forces us to look closely at whether champions of sexual reproduction are hypocritical and inconsistent, allowing one set of reasons for sexual reproduction but demanding another for asexual reproduction.

In talking around the country about cloning and in reading the voluminous literature about it, I am impressed with how many people see themselves as experts about the psychological good of children. Meanwhile, psychiatrists, anyone with a Ph.D. in psychology (even experimental), ministers, counselors with degrees in education, theologians, social workers, licensed clinical psychologists, and pediatricians seem to believe that their experience proves that gay men or lesbians make bad parents, unmarried couples make worse parents than married ones, interracial couples create more stress for children than same-race couples, children are traumatized to learn they were adopted or were twins separated at birth—or were originated by in vitro fertilization.

When I hear these frequent claims, I immediately ask, "Where is the evidence to back that up?" Although the professional making the claim can rarely cite any study, that does not alter his belief that he is correct. If a study is cited to the contrary, say, that gay men make just as good adoptive parents as heterosexual couples, the evidence is discounted.[6] It is a nonfalsifiable claim.

Many similar claims are nonfalsifiable in principle. How do we know that children told of their special origins will be harmed by this knowledge? Even if one or two anecdotal cases are cited, how do we know they aren't the sensational, aberrant case that attracts the media?

To prove the point, identical twins separated at birth would have to be put in an experiment where one is told and the other is not. First, it is difficult to imagine any institutional review board approving such a lifelong experiment. Second, wouldn't it be unethical not to tell each twin of the other's existence? It would, by depriving the child of a possible good friend in her life. For both of these reasons, such studies will never be done, so real evidence will never be obtained.

One way to evaluate the argument that cloned children would be harmed by being created according to traits chosen by parents is to subject the argument to the following thought experiment derived from the idea of reincarnation. Suppose I could choose a genotype that would be mine in my next life (of course, I wouldn't remember anything of my previous life) or I could choose a random mix of genes of the kind that occurs in sexual reproduction. Would the reader leave it to chance or choose a particular person's genotype to re-create?

Furthermore, in this thought experiment, does the imaginative reader think he or she would be harmed by such a choice? I doubt it. Indeed, compared to the randomness of sexual reproduction, I daresay that most readers would actually be benefited by having this choice and having a predictable set of genes for their base in their next life.

Similarly, having birthed or raised a child, every parent seems to think he or she is an expert on children and parenting. Like sex among teenagers, having done it once makes you an expert. Even though some people are lousy parents, it is precisely these people who often think of themselves as the exemplars of proper parenting. No bad parent ever thinks of himself as a bad parent; each is an instant expert on whether a child originated a different way would be harmed by knowing of his unique origins.

Psychological claims of this kind are surprisingly resistant to contradictory evidence. Perhaps because babies and children are perceived as innocent and vulnerable, we feel strongly that they should be protected. Such beliefs probably arose during our evolution and are now hardwired into us. But we shouldn't protect babies so much that we don't create them or ban new ways of creating them.

To conclude, most of the criticism of human cloning based on alleged psychological harm to the child is wildly speculative and seems more a projection than fact. When origination of children by cloning becomes physically safe—when it is no more dangerous than sexual reproduction—we should not let worries about psychological harm to cloned children carry much weight in public policy.

In the meantime, wild speculation about psychological harm to cloned children, combined with making such origination a federal crime, is likely to create a self-fulfilling prophecy. Because we expect cloned children to be odd, we treat them as if they were odd, in effect making them odd. It is like parents who resist interracial marriage because they know that such marriages have too many problems; their own prejudicial attitude helps create those very problems when the opposite attitude could smooth them away.

Indeed, the entire argument about future psychological harm to cloned children reeks of perverting prejudice into ethical justification: because most people fear or loathe something, such fear and loathing is cited as why we shouldn't do it. On those grounds, we should reject gay and lesbian lifestyles, interracial marriages, vegetarianism, religious freedom for atheists, and many necessary procedures in medicine that strike first-time patients as repulsive. All this is a little nuts.

7

Deciphering Cloning at the Earliest Stages of Life

Are religious groups justified, solely on the basis of their theological and religious views, in seeking to control the direction [in America] of federally funded biomedical research?

—*Dartmouth Professor Ronald M. Green,* Human Embryo Research Debates

In the years since the announcement of Dolly's cloning, there has been an avalanche of news about stem cells, cloned human embryos, adult stem cells, stem cells from cloned human embryos, mixed human–rabbit or human–cow embryos, stem cell lines, research embryos, spare embryos, master genes in stem cells, creating eggs or sperm from stem cells, and how all of the above may be related to curing human diseases or reproductive human cloning. Is it any wonder that people who have not taken a recent course in embryology find all this bewildering? Or that they find it nearly impossible to separate hype from fact?

The following discussion summarizes these developments, attempts to make some sense of them, and analyzes their ethical, religious, and political implications.

CLONING IS CLONING

"Clone" is derived from the Greek *klon* for twig or shoot. Most generally, cloning asexually produces a genetically identical copy of an ancestor. Agricultural scientists asexually produce genetically identical copies of prize flowers, fruits, and vegetables.

In the life sciences, "cloning" has several meanings that cover molecular cloning, cellular cloning, embryo twinning, and somatic cell nuclear transfer (SCNT). In molecular cloning, strings of DNA-containing genes are duplicated in a host bacterium. In cellular cloning, copies of a cell are made and result in a *cell line*, a very repeatable procedure that can grow identical copies of the original cell indefinitely. Stories in the news frequently discuss stem cell lines.

In embryo twinning, an embryo that has already been formed sexually is split into two identical halves. Theoretically, this process could continue indefinitely, but in practice only a limited number of embryos can be twinned and retwinned. This use of "cloning" is dying out.

Finally, there is the process of taking the nucleus of an adult cell and implanting it in an egg cell from which the nucleus has been removed. This is popularly called "cloning" but is more precisely known as somatic cell nuclear transfer. If the goal is to produce an embryo that will not be implanted in a woman's uterus, the process is called *embryonic cloning*. If the goal is to produce an embryo to create a child, it is known as *reproductive cloning*.

The process of creating the cloned embryo is the same in embryonic and reproductive cloning. People who try to distinguish between the two are manipulating the facts for political purposes: the difference is limited to the embryo's destination, not its unique, asexual method of creation (more on this argument in the following chapter).

A TIMELINE OF DEVELOPMENTS

In July 1996 PPL Therapeutics had applied for patents from the English government and seven months later, royalties from patents having been secured, it announced the birth of Dolly on February 24, 1997. A lot has happened in embryology and the early life sciences since that day in February 1997. In December 1998 immortalized human stem cell lines were created by John Gearhart of Johns Hopkins University and James Thomson of the University of Wisconsin. Gearhart started with tissue from aborted fetuses; Thomson dissected human embryos left over from assisted reproduction. Gearhart and Thomson discovered how to create a biological factory for continually producing stem cells instead of tediously deriving minute amounts from embryos and fetuses. For this achievement, they will probably one day be awarded a Nobel Prize.

THE IMPORTANCE OF STEM CELLS

What are stem cells and why are they important? Stem cells are primordial cells that have the potential to develop into any kind of differentiated cellular tissue: bone, muscle, nerve, and so on. In theory, they could be grown into specific cells for almost any part of the human body. Found in embryos and the umbilical cord, stem cells also help us grow new cells when we are injured. If we are hurt and lose blood, stem cells are activated to make new blood. If these primordial cells could be controlled by scientists, in theory they could be directed to form new bones, neural cells, cardiac tissue, and hence help cure many diseases.

In 2003 doctors in Michigan injected a boy with stem cells from his own blood after he was shot in the heart with an automatic nail gun.[1] Because it was considered an emergency procedure to save the boy's life, the physicians did not seek FDA approval, although later the FDA told the physicians not to repeat it. The procedure apparently succeeded, since six months later, the boy had recovered much faster than expected.

Other researchers have injected adult bone marrow stem cells directly into the hearts of fourteen people with advanced cardiac failure and apparently helped them, motivating German researchers to successfully repeat this experiment successfully with five of six patients.[2] Injections of stem cells have also been used to dramatically increase chances of survival for patients with multiple myeloma, a cancer of the blood.[3]

A crucial point concerns rejection of foreign tissue by the immune system. Stem cells from existing human embryos will not match a patient nearly as well as stem cells made from human embryos created from a patient's own sperm or egg. Because many sick patients are elderly or no longer produce viable sperm and eggs, creation of such human embryos by the asexual techniques of cloning will be especially productive in producing tissue-matching stem cells for sick patients. The downside is that the need for such precisely made stem cells involves not one but two major ethical controversies of our time: creation of human embryos as means to some other end than birth and use of cloning to create human embryos.

Gearhart and Thomson made their discoveries using private funds. Medical researchers immediately understood the significance of their achievement, but then a larger issue loomed: given that the National Institutes of Health is the treasure chest for the world's scientific talent, should such a chest be opened to fund this new research? (More soon on the politics of this issue.)

On November 12, 1998, Jose Cibelli of Advanced Cell Technology of Massachusetts (now a professor at Michigan State University) announced making differentiated human cells revert to a primordial, pluripotent state by fusing them with cow eggs. In fusion, researchers put the donor cell next to an enucleated egg and fuse the two with a tiny electric current. A blastocyst (or blastomere), a preembryo of a hundred cells or less, starts to develop because the pulse that produces fusion activates egg development. In fusion, mitochondria (primitive organelles carrying a small number of genes) from both the donor and the egg recipient mix, whereas in strict transfer of a nucleus by SCNT, mitochondria are only present in the enucleated egg.[4]

Cibelli made human stem cells out of normal human cells using a cow's egg as his Petri dish. The cow's egg, from which the nucleus was removed, was the medium in which the nucleus of the human cell developed. This was not a mixed breeding experiment to produce something like a mule or, worse, centaur. Nevertheless, the procedure worried critics who sounded alarms about human–cow hybrids. President Bill Clinton and National Bioethics Advisory Commission (NBAC) immediately condemned any attempts to create children out of such hybrids (although no one was planning to try it).

For several reasons, most scientists scoffed at ACT's results. First, the mixed embryo was a dud and only developed to six cells, and even then it wasn't clear whether Cibelli had actually created a cloned embryo. Second, the results were published in an online journal of dubious quality. Third, the announcement of the result seemed timed to raise money for ACT, especially when CEO Michael West appeared on NBC's *Meet the Press* the day of the announcement.

In March 2001, in a development that went under the radar screen of most Americans, researchers at the Institute for Reproductive Medicine and Science of Saint Barnabas in Livingston, New Jersey, created healthy eggs by injecting healthy cytoplasm (ooplasm) from eggs of healthy women into the unhealthy eggs of women who were unable to become pregnant because their eggs lacked healthy cytoplasm. The transfer injected a small amount of mitochondrial genes from the donor eggs into the recipient eggs. When sperm combined with the newly healthy eggs, embryos were created, ten of which became healthy children.

Although in reality just a neat trick to help infertile woman conceive (which Jamie Grifo pioneered and Chinese researchers adopted with Grifo's approval), critics went ballistic when researchers claimed they had done a stealth germ line modification on human beings, meaning that the changes in genes would be inherited by the children of the ten children.[5]

Although it had condemned reproductive cloning three years before, in June 2001 the National Bioethics Advisory Commission concluded that the government should fund research on stem cells created from human embryos. That recommendation was never accepted by Congress. On August 11, 2001, President George W. Bush rejected use of federal money to fund research involving human embryos.

In the literature of medicine and medical ethics, embryos that are unused after successful in vitro fertilization are called spare embryos. In contrast, human embryos created to be used in experimentation are called research embryos. Creation of research embryos carries more political heat than creation of spare embryos. President Bush rejected use of federal funds to create research embryos but said he would fund research on sixty stem cell lines created from spare embryos. He announced this on television in prime time, signaling a new emphasis in domestic government policy on bioethics.

Two years later, the actual number of sources of stem cells is small, less than a dozen, and scientists say they haven't gotten the stem cells they need. Nevertheless, the high-profile television address by President Bush upped the ante for all sides and made embryonic cloning one of the hottest nonterrorist issues in the country. Later, the intensity of this debate was reflected in Congress: "those who are interested in values" should vote to ban therapeutic cloning, said House majority whip Tom DeLay of Texas, calling such cloning "monstrous science." Representative Christopher Smith of New Jersey agreed: "Cloning human embryos for research purposes is unethical, it is wrong, and it ought to be made illegal."[6]

On January 25, 2002, about three years after Thomson's and Gearhart's discovery, Catherine Verfaillie of the University of Minnesota claimed to discover that general, pluripotent stem cells existed not only in human embryos but also in human bone marrow.[7] The newly discovered multipotent adult progenitor cells (MAPCs) appeared to have the same properties as embryonic stem cells (ESCs).

Here is the account of MIT biologist Robert Weinberg:

> Many of our tissues are continually jettisoning old, worn-out cells and replacing them with freshly minted ones. The process depends on a cadre of stem cells residing in each type of tissue and specific to that type of tissue. When an adult stem cell divides, one of its two daughters becomes a precursor of a specialized worker cell, able to help replenish the pool of worker cells that may have been damaged through injury or long-term use. The other remains a stem cell like its mother, thus ensuring that the population of stem cells in the tissue is never depleted.
>
> Until two years ago the dogma among biologists was that stem cells in the bone marrow spawned only blood, those in the liver spawned only hepatocytes, and those in the brain spawned only neurons—in other words, each of our tissues had only its own cadre of stem cells for upkeep.
>
> Once again we appear to have been wrong. There is mounting evidence that the body contains some rather unspecialized stem cells [MAPCs], which wander around ready to help many sorts of tissue regenerate their worker cells.[8]

Verfaille's discovery raised the tantalizing discovery that all the benefits of research with stem cells might be available without raising the abortion-related controversies surrounding ESCs. People immediately chose sides, touting the parity of MAPCs and ESCs or the superiority of ESCs.

Two years later, a growing sense of unease clouded Verfaille's claims about multipotent adult stem cells. In particular, scientists at Stanford University who tried to replicate Verfaille's work found that adult stem cells from bone marrow merely *fused* with existing cells to grow new tissues, instead of becoming new types of cells. One study showed transplanted bone marrow cells fusing with brain cells, not undergoing a kind of metamorphosis, and another found them fusing with cells in the heart, brain, and liver. As one summary puts it:

> Two studies published earlier this year cast doubt over these results—they showed that transplanted bone marrow cells (MAPCs) in mice were simply fusing with existing liver cells, rather than turning into new liver cells. At the time, some scientists thought that cell fusion might be peculiar to the liver, but the two studies published last week show that the same thing may be happening in other tissues.[9]

As this book goes to press, doubt exists not only whether MAPCs can function as well as ESCs in medical research, but even whether they exist at all.

On May 2, 2003, researchers at the University of Pennsylvania made the startling announcement that stem cells can be made into egg cells, which previously were thought to be primordial and inherited in some ultimate way. Researchers created mouse ooyctes from mouse embryonic stem cells. James Batty, chairman of the stem cell task force for the NIH (National Institutes of Health), described the discovery as a "spectacular piece of science." Moreover, such ooyctes appeared to be able to develop parthenotically (without sperm) into embryos.

The discovery touched several ethical controversies. First, and with a longer history, conceptionists have held for over a century that the union of egg and sperm to create a human embryo marks something special and unique; hence, the special status of personhood begins here. They oppose an emergency contraception called Plan B because it prevents an embryo, which they regard as a person, from implanting, or it dislodges one. But if billions of eggs and embryos can be created mechanically in the lab, where is the special status of personhood? And this is what embryonic cloning potentially allows scientists to do.

Second, critics of research on embryonic stem cells, such as Wesley Smith, have recently argued that practical medical research of benefit to patients might require thousands of eggs from young women to grow the required embryos and tissue. Because such eggs are difficult to retrieve (and cost $20,000 for a dozen eggs), large-scale research creates the unsavory image of thousands of poor young women being employed to supply eggs for researchers funded by NIH. The risks posed by drugs needed for such continual egg creation might be detrimental to the donors. Research at the University of Pennsylvania, as well as that by Israeli researchers creating egg follicles from aborted human fetuses, indicates that such a nightmarish scenario need not occur.[10]

Perfectly sensible compromises exist that would move science forward and sacrifice nothing of value. For example, British researchers Daniel Brison and Brian Lieberman of St. Mary's Hospital in Manchester propose that they be allowed to create research embryos from leftover eggs obtained from in vitro fertilization. Women attempting IVF are usually given drugs to induce superovulation and "for every 10 eggs collected in a cycle, three or four do not fertilize and are routinely discarded. In our unit alone, this adds up to 2000 eggs per year and, in the UK as a whole, perhaps 50–100,000. If these were matured in vitro and/or injected with sperm from a fertile donor using intracytoplasmic sperm injection (ICSI), a large number of viable embryos would result."[11] Critics who think embryonic cloning would require thousands of young women to become egg donors are incorrect: the eggs already exist and are being wasted. All we need to do is change our attitude and harvest them.

Third, even male cells can be made to make egg cells, meaning that it might be possible for a gay male couple to create children by mixing sperm from one of the men with egg cells created from the cells of the other, not by cloning the adult, differentiated cells of one of the men.[12] In a related discovery, researchers at the Whitehead Institute in Cambridge, Massachusetts, created sperm from germ cells and then fertilized eggs from mice to create embryos.[13] Combined with creating eggs, this is one more step to creating the building blocks of life from much simpler stuff.

On May 30, 2003, a master gene in stem cells was discovered. Before this discovery, the mechanism by which embryonic stem cells functioned to become almost any cell was unknown. On the above date, scientists at the University of Edinburgh and Japan discovered a gene that allows ESCs to do just that.[14] They named the gene "Nanog" after a fantasy land in Celtic mythology where residents are forever young.

The discovery raised the theoretical possibility that in future years ESCs can be made from adult cells told to revert to embryonic stem cells. Studies to do that, of course, would require more research on embryos and, ideally, federal funding. Although such research might one day reduce the need for research on embryos for stem cells, Richard Doerflinger of the U.S. Conference of Catholic Bishops sternly opposed it, arguing that the ends do not justify the means.[15]

In a related discovery, scientists at the University of Wisconsin learned how to knock out or delete a gene from mouse embryonic stem cells.[16] The ability to add or subtract genes in such a way is thought to be key to telling such cells to develop into a particular kind of cell, such as cardiac cells. It also might bypass the need for human embryos, as the specialized cells could be made directly from the stem cells. Again, studies to do this in humans would require human embryos for research. They cannot be done on mouse embryos or by computer modeling.

WHAT IS AN EMBRYO?

Because of controversial research on live-born fetuses and a general opposition to abortion, Congress in 1977 essentially banned federal funds involving research on human fetuses or embryos. Although a number of panels and commissions since then recommended various ways that the National Institutes of Health and other government agencies might fund research involving human embryos, no Congress during the Reagan, Bush I, Clinton, or Bush II administrations overturned this ban. When President George W. Bush proposed research on a few stem cell lines created from existing embryos, he was cracking this door a tiny bit to allow some research.

So given this controversy over research on human embryos, Americans could be forgiven if they assumed that embryos were easy to define. In fact, that is not true. Dartmouth bioethicist Ronald Green, who has worked on the NIH Human Embryo Research Panel, explains in his book that whereas people formerly defined embryos as being created at the instant that sperm penetrated egg, we now know different.[17] Eggs emit chemical signals that attract sperm, and sometimes several sperm penetrate the egg's outer membrane at the same time. With penetration of this outer membrane, lots of changes start, none of which inexorably results in one unique embryo: the earliest stages during the first few cell divisions are caused by chromosomes in the egg, which explains why parthenogenesis (asexual duplication) can be achieved where an egg begins to form an embryo by electrical or chemical stimulation (this stops after four or eight cells are created). But in these early stages, a parthenotic human egg looks and acts like a sexually created human embryo.

Before this point, and after penetration of the egg's outer wall, the egg sheds half its chromosomes and pushes the remaining twenty-three to the cell's center, where syngamy (the union of two sets of chromosomes) will occur about a day later with twenty-three chromosomes from the sperm.

Even at syngamy, a distinctive embryo does not seem to be present, for there is no nucleus yet. What some call the "zygote," in a day to a day and a half after sperm penetration, grows into a two-cell entity that has a nucleus. Which stage of this process, then, may be studied with federal funds and which is banned? Study of sperm and eggs alone presumably is legal with federal funds. Could it be legal to do such studies before there is a nucleus at the two-cell stage? Possibly.

Instead of asking what constitutes an embryo, suppose we asked what characterizes the *lack* of one. For example, if an embryo-like thing developed parthenotically from an egg but never reached the blastula stage and could never become a fetus, is it sufficiently not special that research could be done on it? We know that genetically abnormal embryos cannot implant in the uterus and are lost in sexual reproduction, which happens to about half of embryos conceived. If such embryos are "born dying," can they be subjects of research, or does this sound too much like research on babies who are born dying?[18]

DO EMBRYOS HAVE DIGNITY?

In Germany, France, Belgium, and Holland, where less than 2 percent of people regularly attend church services, defending the embryo against medical research cannot be based on appeals to the sanctity of life because that sounds too religious and too much like the Catholic Church is calling the shots. So in Europe the same work is done by appealing to the dignity of human life.

Leon Kass borrowed this European concept of dignity to do the same work in America.[19] Kass has failed miserably in doing so, in part because his natural supporters appeal to sanctity of life and its religious connotations, in part because dignity is impossibly vague and cannot do the work that Kass wants it to do. I believe that this use of dignity is religious and equivalent to asserting the sanctity of human life from conception. To see this, we must understand that dignity is properly understood not as a property that an embryo or a human can possess. To think so is to commit another category mistake (see below).

Instead, human dignity *is a way that beings treat one another*—dignified, undignified, or neutral. If dignity were an inherent human property, an individual could not lose it, even if she behaved in undignified ways all the time and other humans treated her as if she did.

What British philosopher Gilbert Ryle called a category error occurs here.[20] This happens when an item in one logical category is mistakenly confused with one in another, for example, when a naive person asks to see a marching band's "team spirit." So too, we know what it means to say that an adult human being must be treated with dignity, or that the elderly must be so treated, although it becomes unclear how a child or baby can be so treated. Although we know how children can be treated badly or immorally, children and babies by nature behave in undignified ways and rarely insist on being treated in dignified ways or even want to be so treated.

New York bioethicist Ruth Macklin argues that "dignity is a useless concept in medical ethics and can be eliminated without any loss of content."[21] She specifically cites uses of the term by the Nuffield Council on Bioethics about genetics and the (American) President's Council on Bioethics about cloning as amounting to vague restatements of other arguments or mere slogans.

The document circulated by the Holy See goes on to argue that cloned human embryos "enjoy the same dignity proper to every human being."[23] That dignity ultimately is a religious concept is made clear, paradoxically, when the Vatican so vehemently denies that its appeal to dignity is "a religious one."[22] While so denying, it then asserts that human liberty cannot encompass the freedom to destroy the "life itself" of the human embryo. In arguing against embryonic research, it says, "How many human lives are we willing to take in this process?" So dignity collapses to sactity of life in the Vatican position.

John Haas, president of the Pope John Center for Ethics in Health Care, testified about cloning before Congress and made the Catholic position very explicit: "A federal ban against the attempted cloning of human beings would certainly be consonant with Catholic moral teachings. But it must be an honest ban. Human life must be protected from its very beginnings, as soon as there is interior, spontaneous growth."[24] Haas appeals not to the sanctity of human life but to its dignity, even in its most nascent forms. For this reason, "opposition to cloning [must] include engendering human life for any research or experimental purposes." Similarly for Leon Kass, cloning human embryos for research threatens human dignity because it transforms "procreation into a form of manufacture."[25]

This position on dignity figured highly in the attempt by a coalition of countries in the United Nations, led by the United States and Costa Rica (fronting for the Vatican), to ban all forms of cloning: "Every human being has intrinsic dignity and worth from conception to natural death," said Birhanemeskel Abebe, an Ethiopian delegate to the United Nations designated as point man in the attempt to get a U.N.-sponsored worldwide ban on all forms of cloning.[26]

When it comes to treating human embryos with dignity, the concept nearly collapses. As UCLA bioethicist Greg Stock argues, "To claim that legislation that so elevates the status of a pinprick of cells that it blocks research to cure real disease afflicting real people and destroying real lives—to call that respect for human life and dignity is absurd. It's wrong."[27]

Perhaps some miniscule sense can be made of embryonic dignity: not eating them, not making them into earrings, or not putting them in plastic for tokens on key rings. But even if it were undignified to do so, even if it violated "the dignity of the embryo," *who would be harmed by doing so*? The stock answer is that it would cheapen human life in general to do so. And of course, the claim is completely unverifiable or unfalsifiable, expressing a sentiment on the part of the speaker that it feels wrong.

But logic sometimes defeats emotion. Consider that nearly 400,000 human embryos now exist in storage, deteriorating each year until after a decade they are no longer viable.[28] Does this practice cheapen human life? How? Considering that very few people even know about it, how could it possibly affect most people? Exactly the same situation would occur with medical research on human embryos: few people would see it, experience it, or even know about it. Its effect on public morals would be minimal, if any. In contrast, the effect of putting embryos on key chains or earrings would affect many people immediately, offend them mightily, and create some plausible argument for a detrimental effect. But at best this only shows that we should ban eating or displaying embryos, so as not to unnecessarily offend people, not that we should ban medical research on them.

Finally, where is the argument that medical research on embryos cannot be compatible with their dignity? Critics say you cannot do research on an embryo if ultimately you mean to destroy it, but that argument makes "dignity" collapse into "sanctity of life," where both mean that no human life can ever be destroyed for any reason. Can you say "circular reasoning"?

Furthermore, given that embryos can be created from eggs in turn created from adult stem cells, the concept of the dignity of the embryo begins to collapse into the concept of the dignity of the cell. Do we want to go that far? Given our above discussion about the difficulty of precisely defining the embryo or its death, can we really give much weight to the concept of the dignity of the human embryo?

Moreover, given the frequent charge that embryonic cloning cheapens life, it is amazing that the parents of many embryos refuse to pay anything, or do anything, for their upkeep and continued existence. An article in the

British medical journal *Lancet* in 2000 revealed that over two-thirds of the human embryos stored at two fertility clinics in England had to be destroyed because the couples whose gametes were used to create the embryos would not even respond to a letter asking about their wishes.[29] The rate of response was twice as high in private clinics that charged the couples fees for continued storage, versus public clinics that did not. In England, the law allows such embryos to be destroyed after five years unless written consent is obtained for preservation for another five years (with attendant costs in private clinics).

Given that such couples are the most affected by the destruction of the embryos and have the ability to give them to other couples for use in conception, it seems clear that they do not put much value in the dignity of these embryos or at least when it comes to paying money to keep them alive or writing a letter consenting to keep them alive in public clinics.

The fact that 400,000 frozen embryos are deteriorating over time and eventually become nonviable has created a new kind of adoption: embryo adoption. The Snowflake program has arranged adoptions of nearly a thousand embryos, about twenty of which became babies.[30] The program charges substantial adoption fees for this service.

England allows scientists to create human embryos for research and use them for up to fourteen days of development.[31] No great changes in the fabric of English life seem to have occurred as a result of this law, nor has there been a massive slide down any greased slippery slope: children are not sold in Piccadilly Circus and homeless beggars have not been kidnapped and killed for their organs. Despite the thunder and vituperation from bio-Luddites, the effect of this law on English life amounts to little or nothing.

LEGALLY PROTECTING EMBRYOS VERSUS REPRODUCTIVE RIGHTS

Either the embryo has protected legal status or it does not. To say that embryos are not persons and can be killed in private but cannot be studied and killed in research is contradictory. Perhaps this shows once again that the cloning debate is really about abortion. Certainly embryonic cloning appears to be so. Isn't it crazy that a woman can abort a sexually created embryo by doubling up on birth control pills but can't create such an embryo as a scientist for research? Or donate it to medical research instead of destroying it with birth control pills?

THE SOUTH KOREAN BREAKTHROUGH

On February 12, 2004, just eleven days shy of the sixth anniversary of the announcement of Dolly's cloning, South Korean scientists announced that they had successfully cloned viable human embryos and also derived viable stem cells from these embryos.[32] Woo Suk Hwang and Dr. Shin Yong Moon of Korean's Seoul National University created 213 embryos and grew 30 of them to the blastocyst stage of about one hundred cells. This is significant because researchers at Advanced Cell Technology in Massachusetts were never able to grow human embryos beyond a few cells, and because blastocysts contain an inner mass of stem cells that have the capacity to form every kind of specialized cell.

Indeed, critics such as MIT's Rudolf Jaenisch had previously implied that it couldn't be done because the eggs were too fragile (perhaps explaining why the experiments at ACT had failed), and if done, the cloned embryos would be abnormal and not suitable as creators of stem cell lines.[33] Like previous claims that cloning animals violated a law of nature, these claims seem to have been proved false.

The well-documented South Korean study, carefully presented and reviewed before publication in *Science*, a leading scientific journal, showed how such an announcement should be done and vividly contrasted to the many claims that they had done the same—always without any evidence—by Panayiotis Zavos, Severino Antinori, and the Raelian sect.

At the meeting of the American Association for the Advancement of Science in Seattle where the announcement was made, the leading ethical issue concerned whether the advance would lead to cloned human babies. Several alarmists decried the advance, arguing that it would lead to reproductive cloning. Leon Kass said, "The age of human cloning has apparently arrived: today, cloned blastocysts for research; tomorrow, cloned blastocysts for baby-making."[34] Thomas Murray of the Hastings Institute emphasized that this technique would not "produce a healthy child," as if that was the main worry.[35] "It would be naïve to say this isn't step closer to irresponsible people attempting reproductive cloning," said Gerald Schatten, who failed to clone primate embryos and emerged as a leading critic of human cloning.[36]

A key step for the researchers was that sixteen Korean women volunteered without pay to take hormones for a month to induce superovulation; and researchers obtained 242 human eggs from them. One of the main ethical worries of American critics has been exploitation of such young female egg donors and commercialization of the process of donation. In this case, informed consent was obtained, no women were paid, and the women had the opportunity to change their minds. The researchers thereby obtained a much larger number of eggs than American researchers ever have. For example, Advanced Cell Technology obtained only eleven eggs from a few women and had to pay them $1,000 each.

With such a large number of eggs, researchers were able to try many different techniques, including the successful one, which involved making a microscopic hole in the egg and, in an almost Buddhist trance of meditation that took as long as ten hours, gently squeezing the nucleus out of the egg, leaving some of the major fluids of the egg inside.[37] The successful technique only worked with cumulus cells, which seem especially potent for cloning, and only worked with women.

The Korean researchers eventually got so good that they could grow one in three embryos to the crucial blastocyst stage. Of the final thirty blastocysts (embryos), they were able to get stem cells from twenty of them and grow one of these in a lab dish into a nearly immortal biological engine for producing more stem cells—a stem cell line.

Stanford's Irving Weissman pointed out that one of the most exciting aspects of this breakthrough is that little understood diseases could now be exactly studied.[38] For example, researchers do not understand the molecular genetics of ALS; by cloning tissue from ALS patients, researchers could study exactly what goes wrong and how.

The fact that this development occurred in South Korea and not the United States caused much controversy on both sides. Critics such as Sen. Sam Brownback, feminist Judy Norsigian, and Catholic spokesman Richard Doerflinger called for a federal law to ban such research in the United States, while American researchers such as Jose Cibelli of Michigan State University lamented that they could not obtain equal numbers of such eggs or federal funding.

When this announcement occurred, I was speaking at a reproductive cloning conference in Switzerland and, as an American and a defender of cloning, to a mostly hostile audience. When I remarked as an aside that the South Korean success showed the futility of the Bush administration's efforts in the United Nations to impose its fundamentalist agenda on the planet, I was stunned when the audience spontaneously erupted into loud applause, showing that most of us deeply misunderstand how many educated people on the planet resent American attempts, with weapons, entertainment, and politics, to impose our views on them.

REDUCTIONIST CONTINUUMS

The discovery of the microscope changed a lot of moral views and, in some cases, worldviews. The discovery that tiny life teemed beneath us gave a new perspective on the relation of humans to the cosmos. In developmental biology, the discovery that embryos/fetuses did not jump exponentially at any time in development—at a time that might correspond when God inserted a soul—changed the view of personhood from beginning at quickening to a forced choice between conception (for religious conservatives) and birth (for secularists).

So too with assisted reproduction. The realization that human-discovered and human-created techniques can influence who got pregnant and when, not just the mysterious will of God, changed the view of conception from fate to either chosen pregnancy or infertility as a curable condition. Effort, money, and young eggs now determine who takes home a child, not prayer.

Recent discoveries about the earliest stages of life and how to create it similarly challenge traditional worldviews. These nascent discoveries promise some of the greatest breakthroughs in medicine and human creation. For this very reason, they also promise to be the most controversial. Previously considered fairly intact, now even the concept of an embryo seems fluid, almost arbitrary, especially when huge amounts of political funding depend on its definition. Even the most primordial units of heredity—egg and sperm—now can be created in principle from stem cells from human embryos.

When it comes to reproductive cloning and reductionism, surely bio-Luddites are right to sniff something more than just a radical way of creating babies. And they are right. Surely safe reproductive cloning will make creating human embryos through sex nonunique and less special. Previously the random combination of genes in sexual combination was unpredictable and elevated by romantics to a mystery of creation. When such cloning is safe, creation of children will be not a mystery but a choice, not only if and when but what kind.

Even more radically—if such a thing is possible—such cloning attacks the conceptual centrality of the human embryo in the politics of abortion.

If an embryo is not a discrete, definite thing, but a continuity on both sides, how much more of a diffuse thing is it when one realizes that any cell in a human body can potentially become such an embryo? For that is just what somatic cell nuclear transfer says is possible with an enucleated egg: any human cell (with the right support) can in principle become another human. Surely that makes conception and embryos nonunique and less special. The embryo becomes an in-between point, not a mysterious and special conception.

When we consider cow eggs serving as hosts for growing human embryos, or hybrids with small amounts of nonhuman DNA inserted into resulting children, or mixed-breed hybrids that cross species altogether (see chapter 15), the whole animal world becomes more fluid, with watery boundaries that bleed into one another and seep. Gone is the sense that God or evolution once and for all created rigid boundaries between kinds of beings that can never interact.

Biological science has made these discoveries in part because it is more focused today on smaller particles: genes, proteins, molecules, and their interactions, not ecosystems or whole humans. As such, creation of life begins to seem more a vast continuity of chemicals, molecules, DNA, and RNA, manifesting in different macroscopic structures but all reducible to similar microscopic structures, especially at the ultimate level of physics.

Seen as half empty, such a world is a mess due to scientific arrogance; it is more unstable, more open to change in the future, less a set thing ordained from above. Seen as half full, such a world shows the interconnectedness of all things, a Spinozistic, Hinduistic seamless cosmos of animal-vegetable-molecular life all heaving, growing, and recycling together.

Either worldview is a prism on What There Is. Both worldviews filter What There Is into What There Should Be; both skimp a little on how they make the transitions.

8

Does One Kind of Cloning Beget Another?

But those late twentieth-century scientists made the same mistake as so many of their predecessors. Understanding the true nature of the gene is "beyond the capabilities of mortal man," they said in 1935; it is *impossible* to determine the sequence of the complete human genome, they said in 1974; it is *impossible* to alter specific genes within the embryo, they said in 1984; it is *impossible* to read the genetic information present in single embryonic cells, they said in 1985; it is *impossible* to clone people from adults cells, they said in 1996. All of these impossibilities not only became possible but were accomplished while the early naysayers were still alive.

—*Princeton biologist Lee Silver,* Re-Making Eden: Cloning and Beyond in a Brave New World

Critics of cloning assert that reproductive cloning and embryonic cloning attack human dignity and the sanctity of human life. Leon Kass writes: "All human cloning begins with the same act: the production of a cloned human embryo."[1] He and others believe that cloned embryos will lead into cloned children; as Sen. Mary Landrieu said in April 2003, "Cloning is cloning. That is why it should all be illegal."

Trying to compromise, champions of embryonic cloning argue that cloning cells has nothing to do with reproductive cloning, that even though we should ban reproductive cloning because it may harm babies, we should leave embryonic cloning alone. According to this view, the circus that surrounds reproductive cloning (Dick Seed, the Raelians, Panayiotis Zavos, Severino Antinori) should not deprive science of a valuable new tool, embryonic cloning, also called therapeutic cloning and medical cloning. Reproductive cloning should be sacrificed, these advocates say, perhaps to be championed in a later century. For now we should make it a federal crime to attempt or achieve reproductive cloning. We can also get the United Nations to vote to ban it worldwide.

This compromise preserves the legality of and federal funding for embryonic cloning. "Don't throw the baby out with the bathwater," says the aphorism, but here we want to throw out the baby and keep the more elementary "bathwater."

Although these advocates hope their compromise wins the battle, I do not think they can prevail. Consider how, at the end of 2002, Stanford University's Irving Weissman announced he would conduct research on human embryos produced asexually; he even said that his research was not "cloning," just messing around a little with special cells.[2] Leon Kass promptly called him a liar, implying that Weissman was deliberately trying to engage in cloning without calling it that. And on this point, Kass was correct. Although everyone in biotech wants to divorce reproductive cloning from research cloning, the move reeks of dishonesty. Why is that?

First, the move is false. As discussed in the preceding chapter, originating a human embryo asexually is origination by cloning. Only the intended destination and use of the embryo differ: as a source of medical research or a source of a human baby. What is identical is what is created and how. Call this the *identity argument.* Senator Landrieu is right: cloning is cloning.

Second, anytime a champion of research cloning tries to argue that embryonic cloning can be separated from reproductive cloning, it's because either it's not cloning or it's unrelated to reproductive cloning. Opponents can counter with the identity argument, making the champion of embryonic cloning defend something false and hence giving anticloning critics the high logical ground.

Third, critics can argue, as Leon Kass has done, that once cloned human embryos are around, some of them might wind up a woman's uterus. Some cloned embryos might inadvertently be implanted in a woman in an assisted reproduction clinic. Indeed, it is probable that such a mix-up might occur, given the well-publicized mix-ups between embryos of black and white couples that resulted in a white couple getting a black baby. (Who knows how many mix-ups have occurred among embryos of parents of the same ethnicity, where parents had no obvious tip-off at birth of a mix-up?)

Of course, even if such a mix-up occurred, such a cloned embryo has little chance of becoming a baby. Because most human embryos fail to implant, at best its chances are 50 percent. Second, because clients in infertility clinics are less likely to conceive in general (otherwise they wouldn't be there) and because embryo implantation is only successful about 20 percent of the time, a cloned embryo's chances of becoming a baby are less than 20 percent. Finally, genetic abnormalities are the main cause of failure to implant by embryos, and critics claim such embryos almost always are genetically abnormal, so only very rare cloned embryos will actually implant and not miscarry.

Nevertheless, it could happen. Given that a tiny chance exists that a cloned embryo might become a baby, the question then becomes: Is preventing such a remote possibility worth paralyzing American medical research? Worth ignoring one of the most exciting discoveries in the history of basic biology of the century?

Linkage arguments force us to bite the bullet. Yes, reproductive and embryonic cloning are linked, both conceptually and practically. So what? Yes, if you have hundreds of thousands of cloned human embryos around, one might end up being mistakenly implanted, having a small chance of successful gestation to birth. So what? Given the enormous benefits of studying cloned human embryos in medical research, isn't taking that risk worth it? Especially given that we tolerate thousands of babies being born every day from women who were on alcohol or cocaine while pregnant with the resulting harm to their babies. And sure, two wrongs don't make a right, but there is the question of scale here: a chance of one (possibly abnormal) cloned baby versus the banal, daily reality of fetal-maternal alcohol syndrome and crack babies. And it is possible that the cloned baby would be normal. Is it also possible that sensationalism about cloning diverts our attention from the real issues around us? (One wonders if Kass's bioethics commission will ever issue a report on something so ordinary as preventing pregnant women from taking drugs?)

Fourth, the position that we should sacrifice reproductive cloning forever and make it illegal magnifies the dangers of reproductive cloning and allows it to be falsely made into the Great Evil. Duke theologian Stanley Hauerwas sees cloning as the lightning rod for new Christian activism: "The very attempt to clone a human being is evil. The assumption that we must do what we can do is fueled by the Promethean desire to be our own creators."[3] An ad hoc group called the Human Cloning Policy Institute attempted to persuade the United Nations to ban reproductive cloning but allow embryonic cloning, calling reproductive cloning "a crime against humanity."[4] It asked the World Court for an opinion from its International Court of Justice, taking that position and declaring that "no individual scientist, corporation, business enterprise, or nation has the moral or legal warrant to clone human beings."

As long as this Great Evil view holds sway, all hope is lost. Scientists are naive to think they can ban reproductive cloning and go ahead with studying embryonic cloning. Why is that? First, linkage arguments start to hurt. If reproductive cloning is terribly evil, then it follows that anything linked to it must be evil. For some people, involuntary sterilization of retarded citizens led to Hitler and the Holocaust. As the latter are evil incarnate, so are practices such as involuntary sterilization that allegedly lead to it.

On the other hand, suppose that the study of embryonic cloning eventually leads to safe reproductive human cloning. Indeed, given the history of medicine, that is just what we should expect. If the primary moral objection to reproductive cloning is that it will likely result in genetic errors in reprogramming, then of course we want research to prevent that kind of problem. But how do we do that? The best way is to see how cloned embryos develop and to study them, gestating them in female chimpanzees, artificial wombs, or human volunteers, then aborting them to see which are normal and which are not, then experimenting to see how to create only normally developing embryo/fetuses. This is exactly what proponents of a total ban on cloning fear.

Wait a minute! critics retort. Some kinds of research—like the research Nazi physicians did on Jewish prisoners—shouldn't ever be done, and cloning research falls in this category. Because we know research with cloned embryos leads to something evil, we shouldn't use cloned embryos to study how the Great Evil can be made easier and safer. If studies of embryonic cloning lead to safer reproductive cloning, then we don't want them.

By parity of reasoning, we also shouldn't allow animal studies of mammalian cloning. After all, a mammal is a mammal is a mammal. Now that we've produced cows, pigs, horses, rats, bantengs, cats, and mice, sooner or later, the rest of the animal kingdom will be cloned, especially to save endangered species. If we don't stop cloning mammals soon, we are eventually going to achieve safe, cloned chimpanzee and gorilla babies, and once primate cloning is perfected, people will try to originate human babies by cloning. So why allow anyone to study animals? Why not ban animal cloning?

Does this reasoning sound crazy? Not to some people. Both Britain's anti-abortion group Life and, across the pond, the Consumer Federation of America argue that animal cloning should be banned because it leads to cloning human embryos and human babies.[5]

More globally, if reproductive cloning cannot be discussed in polite company, it becomes a benchmark of evil, immune from rational argument or criticism. Moreover, anything linked to it must go. Unsurprisingly, this is exactly the way Kass argues: what is wrong about reproductive cloning is what is wrong about cloning human embryos: "The central issue of cloning, however [is not the ethics per se of embryo research], and the primary reason to support a ban or moratorium on all human cloning, is this: it threatens the dignity of human procreation. Concern about this threat should lead us to oppose all cloning, including cloning for research."[6]

This is precisely what the new Puritans of bioetech want. Boston University law professor George Annas says now would be a great time for the United Nations to ban all forms of cloning. (Annas previously argued against all forms of surrogate mothers, just as Leon Kass has always condemned in vitro fertilization.) Professor Annas says the whole world could unite by declaring a new model of human origination to be evil. Then we could progress from there, he says, to other agreements about evil in medicine and biotech. Kass agrees, arguing that his fellow conservatives should leverage the widespread "yuck" reaction to reproductive cloning to a ban on most new biotechnologies, lest we lose our dignity and become posthuman.

What both Annas and Kass foresee is not a periodic reexamination of the safety of reproductive cloning or a lessening of people's fears about cloning, but a move out from the beachhead, such that other kinds of scary, human-dignity threatening innovations can also be attacked and defeated. "But the present danger posed by human cloning is, paradoxically, also a golden opportunity. In a truly unprecedented way, we can strike a blow for the human control of the technological project."[7]

So the new dark ages are upon us. Surprisingly little outcry occurs because their champions keep telling us that the darkening skies are not dark but aglow with moral illumination. Because no one will say a human embryo cannot feel and that a thousand such embryos do not add up in moral value of one real giggling human baby, we see the absurd position embraced in Germany that defending the dignity of human life requires that we never create human embryos for medical research.[8]

(On a related biotechnology front, because no one defends science against its repeated assertions that genetically modified vegetables and cloned livestock are identical to vegetables and livestock created by traditional methods, no one counters the increasingly loud, romantic organic food industry, which has seen its profits rise dramatically as it opposes biotechnology.[9] If this continues, we will be led backward into a world of slow, pure foods that only Prince Charles can afford to eat.)

Predictably, proposals to study failures in animal and human embryonic cloning in order to learn about developmental biology and reproductive cloning are shot down as too controversial to fund.[10] The same thing happened to Robert Edwards in the late 1960s, when English authorities shot down his proposals for funding research into in vitro fertilization on the grounds that it was controversial.[11] University of Pennsylvania bioethicist Arthur Caplan supported a worldwide ban, but only on human reproductive cloning. Similarly, in September 2003, various science academies around the world lobbied the United Nations to ban reproductive cloning but wanted to keep embryonic cloning.

By 2004, the two camps had bogged down, unable to reach a consensus, for all the wrong reasons.[12] Fueled by anti-abortion champions in the Bush administration, the U.S. delegation wanted a total ban on all forms of cloning, whereas the other delegation (led by England) wanted to allow embryonic cloning but ban reproductive cloning. The two camps could not agree on a compromise, so nothing passed, mirroring the stalemate the previous year in the U.S. Senate.

INTERLUDE: REPEATING THE MISTAKES OF THE PAST

It's good that neither ban passed. We should always resist quick, emotional prohibitions on any aspect of medicine or science, whether in Congress or the United Nations, for the past history of such quick prohibitions, led by earnest moralists, is not good. For reproductive cloning, we shouldn't cede such a vast amount of ground forever. We have done so before, when American society ceded the ground about artificial insemination of husband's sperm, and we didn't get help for infertile marriages for another hundred years.[13]

There was a time in the early 1990s when it was not politically correct to suggest that physician-assisted dying should be legalized. When Jack Kevorkian helped over a hundred people die, some people thought end-of-life medicine was out of control. But champions of liberty continued to defend the idea, citing the comfort that the Dutch have with the practice over twenty-five years, and the agony of the people who sought out Kevorkian. Eventually most people realized that even though Kevorkian was not the right man to lead the movement, a bigger issue lay underneath. This is why Oregonians in 1997 voted to permit physicians to prescribe lethal drugs to terminal patients.

Other issues in the past were hobbled by loud, eccentric advocates, such as when H. Barry Jacobs in 1983 proposed buying and selling human organs in a market and tried to set up a company in Virginia to do so. Subsequently a congressional committee headed by Al Gore Jr. wrote a bill, which soon passed into law, making such sales a federal crime. Today, with over four thousand Americans dying each year from lack of organs, everyone is rethinking whether we criminalized incentives too fast for organ transfer. Pennsylvania just passed a law allowing vouchers to be given to families who donate organs of patients declared dead by neurological criteria.

On December 3, 1967, an obscure South African surgeon shocked the world by transplanting a human heart from one person to another. Previously it was believed that such a feat was medically impossible, but Louis Washansky lived eighteen days with the heart taken from the body of accident victim Denise Darvall. The event held heavy symbolism. As with the announcement of the cloning of Dolly the sheep, a media firestorm erupted. *Time* magazine called it the "ultimate operation." Surgeon Christiaan Barnard met with U.S. President Lyndon Johnson and appeared on several American television shows.

Not everyone praised the operation. Nobel Prize–winning surgeon Werner Forssman, who invented cardiac catheterization by experimenting on himself, said the operation was "ghoulish" and criticized the "macabre" scene of cutting out a healthy, beating heart from a dying young woman.[14]

But the rest of the world went berserk, including not only the media but cardiac surgeons. Magazines called 1968 the "year of the transplant" as surgeons all over the planet transplanted 105 hearts. In the same year, surgeons transplanted fifty-five livers. Hype ran so far past reality that predictions bore no resemblance to what patients were actually experiencing. Futurists predicted that soon medicine would create "bionic" men and women. Others said immortality was at hand.

But in truth, patients were dying in agonizing ways. Most transplant patients suffered post-transplant psychosis and a terrible quality of life. Reporters didn't want to understand what was really happening. The basic problem was that immune rejection had not been overcome. The body saw the foreign heart as a threat to be rejected, not as its savior.

In 1969 the Montreal Heart Institute said enough was enough and suspended its heart transplant program. About a year later, Harvard did the same, announcing that everyone should observe a moratorium on heart transplants until more could be learned. Thomas Starzl in Pittsburgh said he would do the same with liver transplants. At the time, only about half of transplanted kidneys functioned after eighteen months. Fueled by saturation coverage in the media, people expected the problem of transplantation to be quickly solved. It didn't turn out that way. Nature can't be forced to give up her secrets to meet media demands.

Discovered in 1971, cyclosporine didn't become available until 1976, when the Swiss pharmaceutical company Sandoz started its clinical trials. (Cyclosporine is a selective immune-suppressor derived from a fungus *Tolypocladium inflatum.*) Seven years later, the FDA approved cyclosporine for use in organ transplants and thus opened the door for the explosion over the next two decades for many kinds of organ transplants. Even today, many people do not realize that cyclosporine, unlike penicillin, is no magic bullet. Many organs are still rejected, and cyclosporine, taken over many years, often causes cancer.

What lessons does this story have for reproductive cloning? The parallels are interesting. First, both the first heart transplant and the cloning of Dolly the sheep were accomplished by obscure scientists working outside of big, well-funded research complexes. Second, both achievements were met with a gigantic media overreaction, which ran far ahead of reality with crazy predictions about the future while ignoring the real problems of patients close at hand. Third, in both cases a seminal breakthrough caused an explosion of research around the world along the same lines and then in new directions. Also, some physicians wanted to bring animal studies to human trial before they understood the basic science of how to prevent bad results.

Fourth, and perhaps most important, no one needed to make organ transplants a federal crime to get physicians to stop. Nor did the United Nations need to pass a resolution calling for a worldwide ban on heart transplants. Kass and company say medicine can only be ethical under the control of the public, but when Kass identifies an extreme position with ethics, he undermines his claim. If ethics cannot be ethics in public policy without such extreme politicization, can it help science progress? Fifth, it is unlikely that hundreds of abnormal, cloned babies will need to be born in order for medicine to put a similar moratorium on human cloning. Even without a birth, a de facto ban now exists in America, and the damages from one malpractice suit, the major compelling mental restraint on the behavior of all American obstetricians, would be enough to ban cloning as a practice in America (for more on the controlling effects of such suits, see chapter 13).

The birth of even one defective cloned baby would continue that de facto ban for decades, if not for a century, and hence the first cloned baby must be normal for the practice to continue. If one baby is born who is originated by cloning is obviously defective in some way, then reproductive cloning will be taboo for the next decade, or maybe (if history repeats itself) the next century.

Like heart transplantation, reproductive cloning needs a breakthrough similar to the discovery of cyclosporine to make it safe enough to try in humans. The recent news from Pittsburgh about the dismal failures with primates shows how far we have to go. It took seven years to discover cyclosporine for transplants and fifteen years to bring it to clinical medicine. It is likely that it will take far less time for reproductive cloning than it took to discover cyclosporine, provided that scientists everywhere are allowed to use federal money to study cloned human embryos and how they develop. Even if federal money does not come, private labs may be able to study this very quickly, just as the assisted-reproduction industry financed its own research out of patient fees during the 1980s and 1990s.[15] Similarly, venture capital and private financing for animal cloning, cures from embryonic cloning, and regenerative medicine may propel research forward in human cloning (unless the zealots go the extreme of banning even private funding of cloned embryo research).

The present state of reproductive cloning can be seen as half full or half empty. I prefer the former and emphasize that in early 1997, it was considered a law of nature in physiology that differentiated cells could not be returned to an undifferentiated state. Now we have not only overturned that false view but have learned how to successfully clone cows, pigs, sheep, mice, bulls, cats, and several kinds of endangered species. We are at a very early time in the study of cloning and I expect great, rapid strides, in part because of commercial interest from the pharmaceutical and livestock industries.

The story of heart transplants cautions us all to go slow in what we promise to infertile couples and patients: science and medicine cannot usually deliver good results on demand, even when people have desperate needs, indeed, even when the lack of such results produces death. It took almost a decade to produce the first effective drug against HIV, even though thousands were dying. Although researchers thought it would take months, it took a decade to find the gene for Huntington's disease after a linkage marker was discovered. That's just the way it is. Mother Nature doesn't give up her secrets easily, and usually we have to wait until the usual alliance of hard work and genius gives us the coveted prize.

LINKAGE TO THE WRONG PEOPLE?

Another way to put all this is to ask whether we should even consider arguing for reproductive cloning in the present political climate if it gives credence to, say, the views of the Raelians. Perhaps we should not defend human cloning because any good arguments will be used by such charlatans for the wrong ends. Should we use good philosophical reasoning to arm the bad guys?

Even though I have argued that human cloning should not be tried until primate studies give us normal medical evidence that it is likely to be safe, such an argument about content at one level supports another argument at the different level of procedure. As San Francisco lawyer and champion of the infertile Mark Eibert argues, it is for parents, not governments or bioethicists, to decide what risks to take in conceiving children, and it's a violation of the liberty of parents to restrain them.[16]

I think a writer in applied ethics needs to consider what the philosopher Daniel Dennett calls the environmental impact of his ideas. To use Dennett's example, if an AIDS researcher discovers that it is possible, by adopting a rigorous and complicated regime, to cleanse the body in some cases of HIV, it is possible that she should not publish this doctrine because it is likely to be misunderstood by some HIV-infected people, who will falsely conclude that it is easy to rid HIV from the body and that HIV is therefore no big threat from risky sex.

While there may be reasons for caution in how such results about HIV are explained, it would be a terrible mistake to not publish the results. Once we become hostage to how hostile critics or uninformed people can misinterpret our work, we will only be able to write boring platitudes or innocuous, overly qualified statements.

One must write the truth as one sees it, even if some people use it to find new arguments for their cause. In the same way, a teacher never knows how some students will use what is taught. After taking my course on the founders of world religions, one brilliant student remarked to me at the end of the course, "You know, it wouldn't be that difficult to start a new religion, would it?" He soon thereafter departed for the American West and I never saw him again.

THE CHINESE EXAMPLE

In October 2003, a front-page story in the *New York Times* announced that Chinese physicians had cloned two human babies whose mothers had miscarried.[17] Sort of. The story and the reactions to it illustrated linkage arguments and problems of making a compromise. As explained in chapter 7, the Chinese created an embryo by mixing a woman's ovum and her husband's sperm in a Petri dish—exactly what happens with in vitro fertilization. They learned of the techniques and used them with the blessing of New York University infertility researcher James Grifo, who pioneered them in America. The Chinese researchers took a nucleus from this eight-celled embryo and transferred it into the enucleated egg of a young female donor.

Because the age of the egg is crucial in the viability of an embryo, researchers hoped this technique might help the original woman gestate her own embryo in her own body. Usually the egg of the young donor woman contributes the genes on the female side, which are combined with the genes in the husband's sperm to form the new embryo. The technique used by the Chinese allows the infertile woman to have a child with her own genes and to gestate it too.

Where's the rub in this? According to *New York Times* reporter Denise Grady, "Critics say the technique is perilously close to human cloning, which has been widely condemned, although there is no proof that it [human cloning] has been done or even seriously tried." Critics fear the new technique for the following reasons. First, suppose you consider the new embryo a person; second, in both the new fertilization technique and in cloning, a nucleus is transferred to an enucleated egg with the intention of gestating a human baby. So both cases involve the ominous-sounding phrase "nuclear transfer technology" (cloning).

In 2001 the Food and Drug Administration (FDA) asserted it had jurisdiction over not only human reproductive cloning but also "nuclear transfer and related research." Researchers would need to submit elaborate paperwork and go through extensive review of ethics committees (institutional review boards) before being granted permission to do such research (resulting in being given an IND number—investigational new drug number). That move by the FDA

> put an end to nuclear transfer work in the United States, Dr. Grifo said. He said the application process—normally followed by drug companies—would be too time-consuming and expensive for most infertility researchers working in clinics and universities. In addition, he said, it seemed to him that the research was so frowned upon that it would be rejected anyway.[18]

This example perfectly illustrates the problem raised by the linkage argument: anything linked to reproductive cloning is condemned. Sensing this, Dr. Grifo gave his techniques and work to the Chinese, who produced pregnancies that unfortunately failed for reasons other than nuclear transfer.

Notice the Grifo/Chinese technique differs from reproductive cloning in two aspects: first, the embryo that will become a child was created sexually (genes of a female and male mixed), not asexually, as in reproductive cloning. Second, the cell from which the transferred nucleus was taken was not a differentiated somatic cell but an undifferentiated cell of an early embryo. Only if someone thinks this clump of undifferentiated cells is a person is there any sense to the idea of a nuclear transfer re-creating a "person."

Because their technique is associated with the dreaded word "cloning," things look bad for the Chinese researchers. Their research is being suspended, in part because of pressure from the George W. Bush administration, which has exalted the human embryo to a sacrosanct entity to appease Protestant fundamentalists, the Catholic Church, and a peck of social conservatives.

Immediately after the Chinese announcement, the Chinese government announced stringent controls over all nuclear transfer research on human embryos. According to the *Wall Street Journal*, "The new rules . . . could help quiet foreign critics who have faulted China in recent years for permitting a range of controversial genetic research. It is likely to be viewed positively by ethicists and conservative politicians in other nations, including the U.S."[19]

So reproductive cloning becomes the third rail in American biopolitics ("Touch it and you die"). As Irving Weissman discovered, any research connected to human cloning makes you a target. All of which wrongly assumes that human reproductive cloning is so evil that it should never occur. And that begs the question.

CONCLUSION

We must resist the view that reproductive cloning is just wrong and only focus on legalizing embryonic cloning. To do so gives up too much, ignores the power of linkage arguments, and permanently bans a method of creation that might one day be safe and (to some) desirable. Given the present political climate, perhaps the best strategy is to stall, to let time pass until we have fought a few more wars around the world and people begin to realize that, compared to Americans coming home in body bags from foreign soil, originating life in a new way from cloned embryos is comparatively no big deal.

9

Why Cloning Will Not Affect Genetic Diversity

An ounce of algebra is worth a ton of verbal argument.

—*Geneticist J. B. S. Haldane, attacking early-twentieth-century eugenics*

A common objection to human cloning runs like this: we shouldn't think about allowing human cloning because it would affect the human gene pool by reducing its diversity. Because we know from evolutionary biology and commercial farming with crops and livestock that genetic diversity is a good thing for protection against disease, we should not allow anything to be done to future humans that would decrease the overall strength of humans to survive unknown future diseases.

More common arguments by critics along the same line claim that the practice of reproductive cloning would affect all humans in the future and would fundamentally alter human evolution. Because we are not wise enough to make such massive changes in human nature or human evolution, we should not do it. All these claims assume that reproductive human cloning really would make such big changes.

One hears this objection frequently from biologists or their students. One also hears it from R. Albert Mohler Jr., professor of Christian theology and president of Southern Baptist Theological Seminary in Louisville, Kentucky. Mohler argues that human "cloning would make possible the eventual desexualization of the human race" and allow for "mass asexual production of identical embryos."[1]

One also hears it from otherwise distinguished jurists, such as Eric Posner and Richard Posner.[2] The Posners argue that allowing infertile couples to clone themselves "could have the radical consequences of eventually eliminating sexual reproduction." I shall return to this claim at the end of the chapter and explain why it is so misguided.

There are two reasons why this objection is mistaken. One is easy to explain, the other, more difficult. The first reason is that the objection seems to assume that humans created by cloning would be mass-produced in millions from only a few ancestors. Perhaps the objector foresees famous movie stars or athletes supplying the genotypes of all humans created by cloning. Or that mass production would produce inferior, faded Xerox copies of the original, as in the movie *Multiplicity.*

But this assumption is false. People originated by cloning would not be mass-produced from a few originals for two reasons. First, as Mark Eibert, a San Francisco lawyer and advocate for the infertile, argues, infertile couples would create most of the demand for cloning. They want a child genetically related to them, so they would choose the genotype of a parent or family member as the source for the new child.

Second, the few parents who did not choose their own genes or a family member as the genetic ancestor of the new child would have thousands of ideal ancestors to choose from as genetic resources, not just a few. Most people want a child of their own ethnic group: Chinese Americans would be unlikely to clone a child from the genes of Michael Jordan or Brad Pitt. If human cloning proceeds along these lines, humans will be originated by cloning from all kinds of people. Because of the practical preferences of infertile parents, cloning will have no effect on the diversity of the gene pool.

The second reason that cloning does not threaten genetic diversity illustrates a lesson about how to think about ethics and genes. The lesson has to do with keeping similar topics together on the same scale. I call this the *scale-to-issue problem.* To understand this lesson, let us backtrack to the beginning of the twentieth century, when ordinary people were first hearing about genes and their importance to human development. Although we associate it with Nazi Germany, the eugenics movement originated in America around 1905 and lasted in America until 1935. The goal of this movement was to elevate humanity through selectively breeding the "most fit" people and discouraging breeding "the unfit." The former was to be accomplished by eugenic marriages; the latter, by sterilization.

This early eugenics movement was discredited in part because it was based on a kind of ethnic racism in which Germans, English, and Scandinavians were seen as the "higher" or "most fit" stock, whereas dark-skinned and Mediterranean peoples were seen as the opposite. Such thinking lay behind the Immigration Restriction Act of 1924, which established quotas by country of origin, severely limiting inflow of people from Africa, Asia, Ireland, Italy, Poland, and other areas deemed undesirable. Because of this act, some people saw the Statue of Liberty only as their ships were turned around to sail back to their native lands.[3]

Another injustice that was perpetrated in the name of eugenics was mandatory sterilization. Although the Nazis led in this area by sterilizing 225,000 "mental defectives," America also did so. By 1941, over 36,000 "feebleminded" Americans had been sterilized, especially in Indiana, California, and Virginia.

These sterilizations assumed that genetic disease was caused by inherited defective genes and that sterilization would eradicate the diseases. But eugenicists then did not understand that most people carry genes for diseases that are not expressed, so-called silent or recessive carriers of non-dominant, recessive genes. Nor did they understand that environmental insults or age may cause chromosomal breakage, creating conditions such as Down syndrome.

Given recessive genes, the number of generations that it might take to completely eradicate a genetic disease would not be a few, as these early eugenicists believed, but many dozen. And that would only be true if they could obtain total control over the sex life of every person carrying such a gene. Although the Nazis may have understood the kind of control needed, Americans certainly did not. Or perhaps they thought that such control only applied to "undesirables" such as the thousands of Chinese laborers who were imported to build railroads in the American West and then sent home.

When upper-class German and Scandinavian women had babies with Down syndrome, eugenicists had to either conclude that some such women were undesirables or revise their thinking about the causes of mental retardation. By 1935 leading geneticists such as Hermann J. Muller were writing that eugenics then was "hopelessly perverted," a cult for "advocates for race and class prejudice, defenders or vested interests of church and State, Fascists, Hitlerites, and reactionaries generally." Another leading geneticist, J. B. S. Haldane, said that "many of the deeds done in America in the name of eugenics are about as much justified by science as were the proceedings of the Inquisition by the Gospels." Moreover, given the new knowledge of population genetics, Haldane quipped, "an ounce of algebra is worth a ton or verbal argument."

What did Haldane mean by this? One of the most famous truths of population genetics is called the law of regression to the mean. This law is relevant to the idea of cloning as a threat to diversity. Let's give this idea a little rope and see if it hangs itself.

The human population of planet Earth now is roughly 6 billion, give or take a few hundred million. For the sake of argument, suppose that several million humans were cloned who were seven feet tall. Over time, would these new tall humans affect the mean height of humanity?

Not at all. First, suppose we make the crazy assumption that the whole planet becomes a dictatorship, and seven-footers are only allowed to marry other seven-footers. Even then, height is likely not to be a dominant set of genes but recessive, so if a couple of seven-footers had four children, most are not likely to be seven feet tall.

Second, we cannot really make the horrible, crazy assumption that the whole world is a coercive dictatorship. Even such a dictatorship cannot possibly control who has sex with whom in the remote rural villages of the planet. Absent such a worldwide dictatorship, the seven-foot giants are going to have sex with those they choose, marry those they choose, and consequently will have their tall genes mixed with those of normal people. Over several generations, the tall genes will be diluted.

This is an example of the law of the regression to the mean in population genetics, which states that in huge populations reproducing over many generations, values will cluster in normal ranges. In other words, abnormal values will be normalized over time. People who are very tall or very short (or very smart or very dumb) will find their distant descendants to be very normal.

Which explains two things: we need not worry, as some social conservatives have done, about a declining gene pool. Nor is it ever going to be possible to create a genetic master race through eugenic marriages or designer babies. In all such attempts, and barring total control over reproduction in an unprecedented governmental intrusion into everyday life, the law of regression to the mean will thwart all attempts to change the mean of the human gene pool.

The scale of human reproduction behind this law is difficult to comprehend at first, but picture 6 billion humans in any one generation reproducing themselves, followed by other generations, such that over the next century, 20–40 billion humans will live on this planet. This is such a vast number of babies born, with even more embryos that fail to implant, that the actions of a few thousand (or even a few million) humans inside this vast vortex is rendered insignificant.

In a way, this truth is reassuring. The human genome is remarkably stable. Nothing much is going to affect it, either in a bad way or in a good way. If this means that eliminating all genetic diseases is going to take longer than we thought, it also means that we need not worry about losing genetic diversity through human cloning or attempts by some dictator to create a master race.

If genetic diversity trumped all other values, we could actually insert a gene creating diversity, as Robert Edwards has argued. Only this way we wouldn't deal with the evolutionary roulette wheel of fate but would design certain kinds of randomness we believe desirable, for example, those missing defective genes.

Finally, the argument sometimes given by biologists, that defective genes must be kept for the benefit of the human gene pool, is reprehensible. It implicitly accepts deliberately inflicting diseases such as Tay-Sachs, cystic fibrosis, or sickle cell anemia on people in case humanity one day in the future needs such genes. If such a thing is to be done, surely the victims should volunteer for such a heroic mission.

More practically, human cloning is never going to be possible on the scale of millions. To create an embryo by cloning, in vitro fertilization (IVF) is necessary. At present and for the past two decades, IVF is not very efficient. Of one hundred couples who try it, at most twenty-five will ever take home a baby. And it is expensive, costing about $8,000 per attempt. Only fifteen states require insurance companies to cover it, and the medical insurance carried by most small companies doesn't provide for it. Of the people who use IVF, only a few would want to try cloning the genes of someone not connected genetically to one of the parents.

On top of all this, sex is too much fun to ever be replaced as the way most humans are created. Evolution has created us with a strong drive for sex and, unless we are very careful, when we have sex, babies result. All these facts assure us that genetic diversity is safe from any threat posed by human cloning. Beyond all this, we need to ask why concerns about genetic diversity and population genetics should have anything to do with everyday morality. Here the question arises about asking questions in ethics at the right level.

In essence, we shouldn't bring issues from a different order of magnitude into discussions of ethics. In this scale-to-issue problem, why should a loving woman and man in 1905 or 2005 suddenly worry whether their children will be eugenic or fit? In considering marriage and children, other than wanting to prevent children with genetic disease, why should they think about genetics at all? Why should they modify their personal ethics in favor of public policy goals? The answer, of course, is that it was silly to think they would do so in 1905 and it's even sillier today. Even if a few enlightened people forsook marriage for eugenic reasons, the effect of their actions on the overall human gene pool would not matter at all.

Similarly today, the effect of actions taken by any particular couples on the human gene pool, diversity, or raising or lowering the average qualities of the human race is nil. If a few thousand couples created designer babies, it would have no effect on the human gene pool. To paraphrase John Stuart Mill, where a couple is likely to act for eugenic reasons in having or not having a child, the couple is likely to do so for the wrong reasons, at the wrong time, and with the wrong results.

Given the insignificance of any couple's reproduction on the human race, it is remarkable how often such considerations are brought into arguments in reproductive ethics. Fears about creating a master race or about the declining gene pool, as well as fears about loss of genetic diversity or designer babies, really shouldn't count as good reasons in such discussions of ethics. They are at the wrong level and concern issues of a vast scale of human actions, not those of a particular couple at one time.

This scale-to-issue problem frequently crops up in discussions of new issues in ethics where people inappropriately bring up considerations of a vast scale into a case or practice affecting a much smaller number of people. In a class I once taught, I explained the case for moral vegetarianism, emphasizing the prevention of suffering to animals raised for food. An undergraduate seriously objected, "But if we didn't eat all the cattle and use all the cows, they would just wander around, clogging the interstates." Again a problem of scale, as she failed to understand how falling demand for beef could eventually reduce the number of calves produced for meat.

It's hard in ethics to elucidate when scale-to-issue problems arise because we often do the reverse and encourage people to consider the universal ramifications of what they do in affirming Kant's principle of universalization: what if everybody acted on that rule of conduct? So Peter Singer asks, "What if everyone drove an SUV?" (bad), and "What if everybody didn't eat meat?" (good).[4]

The answer to the problem is that appeals to worldwide consequences must be relevant to the question at hand. If we're discussing global warming, whether everyone drives an SUV is relevant. If we're discussing how to have babies, the effect of a particular couple on the human gene pool is not. We really don't have to act or think as if every time we endorse a change in practice that everyone in the world will be affected by it. It is indeed arrogant to think so. Whether we legalize gambling is likely to have little or no effect on Asian countries; whether Belgium and Holland legalize prostitution is likely to have no effect on North America. And so on.

So back to the claim by Eric Posner and Richard Posner mentioned at the beginning of this chapter that allowing infertile children to have cloned children could eventually threaten the existence of sexual reproduction in humans. This claim has many scale-to-issue problems, as well as making as many dubious assumptions as early-twentieth-century eugenicists did. Its main problem is the assertion that the actions of a few hundred (or thousand or million—it makes no difference) infertile couples would affect the human genome in its actualization by 6–7 billion humans over the next thousand years.

As Texas law professor John Robertson says, to get there, many silly things must be assumed, among them that infertility in couples is always inherited, that children created by cloning would only want to create their own children by cloning, and that people can clone children faster than they can create them sexually.[5] All these assumptions rest on deeper mistakes that assume the mysterious irrational power of cloning over people created this way or using this technique as parents.

Finally, suppose that creating robust diversity in the human gene pool should be the primary value among all human values. In that case, we should eliminate medicine, for surely medical care allows genetically weak people to have kids who otherwise would have died before the age of childbearing. Or we should require people to submit to a gene analysis and make them mate with members of the opposite sex with widely divergent genes, so a short, rotund New Yorker would be bred with a tall, lean Tutsi.

Of course these proposals are ridiculous. So many other values are so much more important than genetic diversity that no one ever thinks about diversity. All of which is to say that we should move away from thinking of reproductive cloning as a special case, a radical way of originating humans that befuddles our previous moral thinking. Once it becomes safe, it will be another reproductive tool that a small minority of the population will want to use.

10

Marxists, Feminists, and Actors on Cloning

Meanwhile, a coalition of a hundred people and organizations recently sent a letter to Congress expressing their opposition to therapeutic cloning—among them Friends of the Earth, Greenpeace, the Sierra Club, the head of the National Latina Health Organization, and perennial naysayer, Jeremy Rifkin.

—*Robert Weinberg,* Atlantic Monthly, *June 2002*

On a recent visit to a large university I talked about arguments why human cloning (when safe) should not be regarded as a great evil. The campus was noted for its liberal faculty. During the question and answer period after the talk, a prominent Marxist professor of humanities (who looked like one of the guys in ZZ Top) became almost red with rage as he exclaimed: "The whole idea of cloning humans is repugnant to me and I find your attempt to justify it offensive." He was really mad. Afterward I learned that he was accustomed to having his views carry the day among both his peers and his cowed undergraduates.

I asked him if he was pro-choice about abortion and of course he was. I asked him how supporting a parent's right to abort a second-trimester fetus because it had a genetic disease differed from supporting a parent's right to choose a healthy fetus. "And I suppose," I continued, "that you would also support a woman's right to abort a healthy second-trimester fetus," which he did. I concluded, "So you support parental choice against genetic disease and against unwanted, healthy fetuses but not for choice of characteristics?" He rebutted, "Some choices should be banned." I retorted, "So who would do the banning?" He exclaimed, "The federal government!" He went on to ventilate that government could, Libertarian style, make inequalities worse or it could decrease the natural inequalities of fate. "Allowing genetically rich and financially rich parents to choose the genes of their children will certainly increase inequality in America," he finished.

I replied that the state had no business banning new ways of making children; that if he really wanted to attack inequality, he should abolish private schools (like his own university) and make all education public and free. "Why pick on the new reproductive option?" I asked. At this point, he became apoplectic.

Our discussion in public lasted too long, and afterward I realized how many people on the political left oppose parental choice about cloning and genetic traits. Mind you, this opposition had nothing to do with abnormalities or technical aspects of cloning; these liberals were *in principle* against parents choosing gene-based traits of their children.

Whereas the biologists in the audience focused on safety and errors in genetic development, the nonbiologists assumed these problems would be worked out eventually and focused on the social-political consequences. Most of the audience seemed to agree with the Marxist professor's way of thinking. For example, they insisted that we need a theory of distributive justice to allocate any new biotechnology before the new biotech is approved by the federal government.

To me, this confused the Good and the Right. Something can be good and yet be unequally distributed (intelligence, health, beauty). Moreover, something can be *justly* distributed but not *equally* distributed (some may not work as hard as others or may have fewer needs). Finally, something may be justly distributed but bad, for example, death.

The surprising thing to me, especially on campus, is that the Marxist professor didn't think my talk should have occurred at all (a bioethicist on campus had pushed for my talk). Professors on this campus certainly weren't used to hearing all sides. Although their Marxist views benefited from freedom of speech, and they would have been the first to cry out if their views were censored, they had no stomach for defending views on the other side.

Shortly after the talk, cloning activist Randolfe Wicker e-mailed me and asked what I thought of Judy Norsigian coming out against embryonic cloning. I e-mailed Randy back, "Who is Judy Norsigian?" He answered,

"One of the authors of *Our Bodies, Ourselves* in the so-called Boston's Women's Health Collective." In another words, a feminist famous in some circles for being pro-choice about abortion.

What Wicker alerted me to was Judy Norsigian's testimony before a U.S. Senate committee where she passionately opposed allowing parents and women to donate their eggs or produce embryos for therapeutic cloning.[1] Thirty years before, she had led the fight against older women and men, especially older Catholics in the Boston area, who wanted to deny women the choice to abort.

I remember well how the book *Our Bodies, Ourselves* affected young female graduate students and assistant professors in the early 1970s in New York City, where I was a graduate student. It caused nothing short of a revolution in challenging the paternalistic attitude of most gynecologists and physicians toward women. The book's message was, reject paternalism! Take control of your body! Learn about it! Make informed choices!

At that time Norsigian opposed the paternalistic view that women could not make their own choices and needed paternalistic regulation and laws. Thirty years later, she reversed herself, arguing paternalistically that young women do not know what they are doing when they sell their eggs. Now she claimed that women should not be allowed to take drugs to superovulate to produce eggs to create embryos ("too dangerous: we need long-term studies to prove safety").

Norsigian emphasized the claim (discussed in chapter 7) that therapeutic embryo cloning, if implemented widely in medical research, might require millions of eggs. Although obviously there are other ways to get eggs (such as using leftover eggs at IVF clinics), the numbers impressed her. Oddly, similar claims by anti-abortionists about the number of abortions (about a million a year for the past 30 years) never impressed her. What's wrong with this picture: Protecting eggs is more important than protecting second-trimester fetuses?

Judy Norsigian really worried that therapeutic embryo cloning would lead down the slippery slope to designer babies. And who would choose gene-based traits of their babies? Why, those very same women who might later choose to abort a fetus. What's next? Outlawing abortion for the wrong reasons? The veil begins to lift: pro-choice is a very diluted principle: you have the right to an abortion but only if you abort for the right reasons. Otherwise your right to abortion doesn't exist.

Norsigian saw no problem with her position in her testimony before a Senate committee,"Although we are advocates for women's reproductive health and have worked decades to support improvements in contraceptive research and development, we do not believe that cloning and genetically engineered children are extensions of 'reproductive choice.'"

Citing Jeremy Rifkin's legal sidekick, Andrew Kimbrell, Norsigian asserted that allowing parents to make choices about kinds of children amounted to commodification of children. "An unregulated industry in cloned human embryos will likely lead to unacceptable commodification of life." This argument conflates the principle of choice with the fact that, to implement some choices, medical professionals have to be paid. It is like saying that abortions should be legal only if physicians provide them for free.

As for commodification of life, both liberals and conservatives are hypocritical about this, watching aghast as young men and women sell sperm and eggs but looking the other way as bait-and-switch pregnancy counseling outfits work in tandem with adoption agencies, in essence selling the nonaborted healthy baby for as much as $50,000 to an infertile couple.

According to a *Wall Street Journal* study in late 2003, babies are bought for adoption from the following countries for these average prices: Guatemala: $22,000; Bulgaria, $18,000; Ukraine, $14,000; China, $11,000; Russia, $17,000; South Korea, $15,000; Kazakhstan, $14,000; Vietnam, $15,000; Colombia, $10,500; and India, $10,500.[2] In addition, many countries require prospective parents to spend several weeks in the country or to make two trips, adding another $5,000–$10,000 to the cost. Approval by the United States that year of the Hague Convention on International Adoption was expected to dramatically increase these costs.

Is this not commodification of life? Why do liberals and conservatives remain silent about it? Why do they spend so much money on nonsentient embryos and stem cell tissue when babies are being bought and sold all over the world? Because the commodification of adoption gets a pass by the media and by conservatives, when anything connected with biotechnology does not. Why is that?

The answer is the same one that applies to the question, What happened to opposition to letting terminal patients die? And to voluntary euthanasia for the terminally ill? When real families agonize whether to let granddad die or adopt a foreign baby, woe to anyone at the PTA meeting or the office who tells them it's immoral. Much easier to attack unknown scientists in far-off labs who kill embryos/babies.

One is tempted to think that the massive attention of the religious right—Protestant fundamentalists, Catholic bishops, and television evangelists—has less to do with protecting embryos than with thumping a hot button for money and power in American politics. Such an issue has the proven ability to move people, like the issues of gay rights, abortion, prayer in schools, the Ten Commandments in public buildings, and creationism.

What my visits to campus taught me, what Judy Norsigian taught me, is that the political left does not support parental choice. Many thinkers on the left envision a very particular theory of the good life (environmentalism as a secular religion) and identify morality with the quest for that life. On this worldview, abortion is good and accepting infertility is good because both reduce overpopulation and hence help the environment.

Increasingly, those on the political left agree with social conservatives that the state should promote a particular vision of the good life (the feds should ban cutting old-growth forests). They understand that others have a different view of the good life (e.g., family values versus women's health), but they dismiss these views as wrong, fundamentalist, neoconservative, or ignorant.

Nor are these rare examples of the views of social liberals on cloning. Well-known Hollywood actors such as Peter Coyote take it on themselves to warn society of the dangers of allowing parents to make choices about the traits of their children. Jeremy Rifkin, who started out pushing the People's Campaign for Social Justice, has made a career of attacking biotech, focusing on the ultimate evil: globe-spanning corporations.

LINKAGE ARGUMENTS, AGAIN

Part of the reason liberals can oppose cloning has to do with the linkage arguments. Because everyone now agrees that reproductive cloning is evil, it's something around which people of all colors and stripes can unite. And if human embryonic cloning might lead to reproductive cloning, then it too should be banned. In an article in the conservative *National Review* that anticipates Norsigian's testimony, conceptionist Ramesh Ponnuru takes the linkage argument one step further and turns it around: "I think, contrary to the prevailing assumptions, that therapeutic cloning is less defensible than reproductive cloning, because the former involves the killing of a human being and the later does not."[3]

Environmentalists press a different kind of linkage argument: they fear that biotechnology is increasingly being offered as a quick fix to environmental problems such as toxic sludge (bioengineered organisms may transform such sludge into harmless by-products). The paradigm is endangered species, where the only acceptable strategy endorsed by the National Resources Defense Council (NRDC), the darling environmental organization of Hollywood and Southern California liberals, is preservation of environment.

Cloned animals, cloned embryos, cloned cell lines—it's all biotechnology, all a quick fix, all for the profits of those who hold the patents, not for the benefit of the planet or the masses: so goes the antibiotech argument of environmentalists. For the World Wildlife Fund, cloning the banteng is not the answer but part of the problem: it allows people to ignore destruction of habitat.

This example of the purity problem in moral thinking reminds me of a similar problem faced by a colleague at the University of Alabama at Birmingham (UAB), Brad Rodu, who switched people who had failed at giving up cigarettes to smokeless tobacco.[4] Dr. Rodu demonstrated that the change could add years to their life and increase the quality of their remaining years. But his work was denounced by the American Lung Association because he did not advocate, and even undermined, the pure, virtuous approach of quitting cold turkey. Yet there is no doubt that, for the carefully chosen group of people who are hopelessly addicted to nicotine, his approach is correct.

Genetically modified food is not really good for the starving, so this argument goes, because it merely allows another billion people to live, consume more resources, make more pollution, and push the planet beyond its natural limits. For this reason, radical environmentalists passionately believe in birth control for people in developing countries (and create one of the few areas in biotech where they dramatically disagree with Catholic clergy opposed to birth control).

A more subtle way to put this linkage argument is to argue that that animal cloning and therapeutic embryo cloning create the wrong mind-set, one of fantasy rather than that of realistic prevention: it's like offering lung transplants to lifelong smokers. Sometimes this argument is put in terms of the "social meaning" of cloning and biotechnology. Just as cloning the banteng seems to say that we can keep on driving SUVs and not worry about how their emissions create global warming (because biotech will fix that too), so we need not keep our weight and bad cholesterol in check because biotech will save us through a heart transplant, an artificial heart, or an injection of new cardiac cells cloned from embryos created from our own cells for this therapeutic purpose.

Here we get a conceptual pincer movement against biotechnology. On one side are bio-Luddite theists who believe in a natural world that should not be harmed or changed. On the other side are bio-Luddite environmentalists who believe that a similar natural environment evolved to its present point and that not even one species should be lost. Destruction and addition of new species is not the goal of either side. Because biotech would allow creation of new transgenic species, it is almost as threatening as nuclear weapons, maybe more, because the new biotech will be a stealth invasion.

HOW HIGH A BAR?

As a reviewer for publishers and journals in bioethics, I have read many attacks on the idea that parents have a liberty interest in reproductive cloning. Most of these papers attack the core idea of procreative liberty and autonomy and argue that it is not sacrosanct, that other communitarian values can limit it, that parental autonomy is not consistent with family values (where the good of the family comes first, not individual choice or preference), and that procreative liberty is only one value among many other good values.

What these papers teach me is that if any of us had to defend in public an idiosyncratic decision, for example, to pursue a philosophy doctorate rather than go into business, to get married late or divorced early, to have eight children or to have no children, it would not pass muster with these writers.

For the above subjects, all academic and popular writing is moot: no one needs to justify getting married, having kids, or pursuing a strange career. Because we have a country that has made personal liberty not just abstract rhetoric but a real value woven into the seamless details of everyday life, people know that such choices are theirs forever. Try to take them away and watch the firestorm that erupts.

Strangely, though, many people simply refuse to generalize the principles underlying American liberty in individual and family life to new ways of making babies. When it comes to new reproductive technologies or cloning, people seem to imagine that these opportunities can be curtailed without squashing their underlying liberty.

To see how this is not so, suppose that cloning a human embryo becomes something that any high school student can do in a lab with a few chemicals, a microscope, human sperm, and a few eggs. (Suppose we learn how to clone a billion human eggs cheaply and easily.) How exactly does the state prevent such embryos from being gestated? Are we going to go the way of the United States before *Griswold v. Connecticut*, banning physicians from prescribing hormonal birth control and IUDs? If so, it's easy to see how a great intrusion into reproductive liberty will occur.

ENVIRONMENTALISTS AGAINST CLONING

Green liberal Bill McKibben, the husband of novelist Sue Halpern, wrote the pro-environmental book *The End of Nature* (1989). Now he finds himself, as he says in a 2002 op-ed in the *New York Times*, along with his friends at Greenpeace, Earth Island Institute, and Friends of the Earth, compelled to oppose biotechnology, not only as genetically modified food but as reproductive and embryonic cloning.[5]

Like Leon Kass, McKibben has a visceral revulsion against changes in the world that biotechnology is bringing. He argues that if a child were created in part for certain talents and had those talents as a child, that her talent would be less valuable. McKibben believes that the individual effort that makes a talent into an achievement is why we praise winners. But he assumes that less effort of will is required for a genetic talent that is chosen versus a genetic talent that comes by random recombination of parents' genes. In either case, involvement by parents may have much to do with whether the talent is realized. Indeed, as an environmentalist living with his wife in a rural home in the Adirondacks of New York, he admits that when his daughter asks, "What are we going to do today?" he always wants to take her on a hike.

Suppose a gene existed that predisposed a girl to being a naturalist. How would choosing that gene for your daughter differ from trying over fifteen years of her upbringing to make her into the same? Indeed, how does he know that her desire to hike isn't gene based?

From what he writes in his book, if she were unlucky enough to inherit an "anti-hiking, anti-outdoors" genetic package (if this exists), and preferred to read indoors and stay on her computer rather than hike, run, and camp, he would strive mightily to change her. But how do the ethics of that differ from choosing an embryonic package earlier? To return to themes in chapter 2 about what cloning tells us about ourselves, what we fear is the overzealous parent striving too hard to change his daughter to his own expectations. Ethically, how that is done is secondary to whether it's right or wrong to do it.

McKibben might reply that it seems like cheating to choose the disposition. And that is true. Given thousands of years of intuitions about "who we are" and the value of the hard road, some intuitions have to change. It is analogous to those who think that mental well-being must always be won through the hard work of intensive psychotherapy, with lots of self-reflection and feedback from the therapist (with the ultimate form being psychoanalysis). But a more sophisticated knowledge of psychopharmacology has caused many of us to rethink that view, especially for people with bipolar (manic-depressive) disorders, chronic depression, and mood disorders caused by hormonal changes. Drugs such as lithium, Prozac, and the benzodiazepines help people function better (much better than alcohol, which to many people becomes a bad drug and a poison). The important point is not whether such drugs are cheating in winning the prize of happiness but that they work where the hard way often does not work. Without the drugs, many people cannot function. (More on the same about cloning and sports is taken up at the end of chapter 16.)

CONCLUSION

The popular British philosopher Mary Midgley once wrote a book titled *Evolution as a Religion* to expose the fervor with which some people promote evolution.[6] In a similar vein, author Michael Crichton recently gave a speech at the San Francisco Commonwealth Club accusing modern environmentalism of becoming a religion for urban atheists. For both authors, religion assumes and supports passionate, unthinking, emotion-led views. Crichton says about this environmentalism:

> There's an initial Eden, a paradise, a state of grace and unity with nature; there's a fall from grace into a state of pollution as a result of eating

> from the tree of knowledge, and as a result of our actions there is judgment day coming for us all. We are all energy sinners, doomed to die, unless we seek salvation, which is now called sustainability. Sustainability is salvation in the church of the environment. Just as organic food is its communion, so is pesticide-free water that the right people with the right beliefs, imbibe.[7]

When it comes to education and inequality, most liberals hope that human nature is plastic, that deep inequality rests not in the genes but in changeable socioeconomic structures. When Catholics want to think for themselves and choose for themselves, they delude themselves that they are still Catholics, when they have become Protestants. When liberals fear that biotechnology will hurt the environment, women, or human nature, and want the state to stop such changes, they too delude themselves that they are still liberals, for they have become social conservatives.[8]

11

Is Safe Reproductive Cloning Good?

I can envisage a time, not in the foreseeable future but one day, when it became safe or relatively safe to produce human clones . . . and when it might be a remedy for infertility which some people might think worth exploring.

—*Baroness and philosopher Mary Warnock, head of the Warnock Committee on In Vitro Fertilization in Great Britain, July 28, 2002*

A recent news release described how a group of professionals approached the World Court, urging it to make reproductive cloning a crime against humanity.[1] Such a move assumes that, regardless of how safe it becomes, reproductive cloning is not wrong because of its lack of safety at present or because the early stage of science now produces too many abnormalities, but because reproductive cloning is intrinsically wrong in itself.

This is not a scientific claim but a philosophical one. Although made by scientists, it lands in that more general realm known as ordinary morality, as do claims about the conduct of physicians who may want to be judged only by the norms internal to medicine but are still judged by norms of truth telling, decency, respect for persons, and fairness that stem from ordinary morality.

Several common arguments assert that reproductive cloning is intrinsically wrong, each with a slightly different twist and in different words but all amounting to the same claim—regardless of how safe it becomes or how much good it might create for a particular family, reproductive cloning is essentially wrong. These arguments include the following: reproductive cloning is an evil practice in itself; it inherently destroys the dignity of the child created; it is incompatible with the sanctity of life; it is against the will of God; it inherently dehumanizes children and treats them as commodities; it is evil because one kind of being is used as a resource for another.

A different class of arguments does not claim that reproductive cloning is intrinsically wrong but asserts that it is nevertheless wrong because it *indirectly* leads to bad consequences to the child, the couple, or society. Here reproductive cloning is claimed to be wrong because it is psychologically bad in some way for the cloned child or because the genes of the cloned child may have some hidden abnormality that may not be expressed until adulthood.

Or it is claimed that cloning is harmful to the parents who create a child with strong expectations, or it is wrong to start a practice in society where children are not loved in themselves as God's gifts but treated as commodities designed to bioengineering specifications.

Finally, critics argue that reproductive cloning is indirectly wrong because of its indirect, long-term consequences for society: that it leads to decreased genetic diversity in the human gene pool, that it sends the wrong message to the disabled, and that it will eventually send the wrong message to normals that there is something wrong about them.

I have dealt with all the arguments above, either in early chapters of this book or in my previous book on cloning.[2] Now I want to turn to the more difficult task of making the strongest possible argument, once animal studies make it safe to try, *for* reproductive cloning.

For the sake of conceptual clarity, in the following pages I assume that the day has come when all mammals and especially primates can be routinely and safely originated by cloning with a rate of defects no higher than that found in sexual reproduction in the same species. I assume that studies of cloned human embryos have shown that they can be created in the same way, without any more abnormalities than occur in sexual reproduction. All that assumed, what's the case for allowing safe, human reproductive cloning?

CREATING WANTED CHILDREN

The first argument needs a little background, so let's detour a moment and consider the oft heard claim that an adolescent or adult will be traumatized to learn that he or she was adopted, was created by in vitro fertilization, has an unknown twin, was created by insemination of donor sperm, or was originated by cloning. As I have already argued, these claims are highly speculative and mostly express the projections of critics.

These critics have us imaging a future when cloned children spend hours brooding over the fact that their parents willed their existence and consciously thought about whose genotype they would embody. They envision such children agonizing over whether they will live up to their parents' expectations, traumatized by the fear that they will not and trembling lest they lose their parents' love.

How silly! To see why, we need not go far into the future but merely turn the mirror on ourselves, for there is one question close at hand that is foundational to anyone who was born before 1965: were we really wanted by our parents or were we just accidents? Were we consciously wanted, planned, and deliriously sought after, or were we an unintended-but-foreseeable byproduct of having unprotected sex in an age when both contraception and abortion were illegal for all couples, married or not?

First, even if we ask our parents and they swear we were wanted (what would you say to *your* kids?), we will really never know. Second, we don't really care about the answer and probably won't ask our parents, but either way, given all the water under the bridge, the answer is irrelevant.

Nor do we ask our parents whether they really wanted a girl instead of a boy. We don't ask them if they planned to have a child but only years later, once they had their marriage on a better footing or had more money. We don't ask if they thought about adopting instead of conceiving, and if when young they considered a childless lifestyle that involved traveling around the world to exotic locations.

No, we usually don't ask such questions and if we do, our parents are usually a bit uncomfortable (Why are you asking me *that?*). Bonding, affection, and the life of the family have long since carried us beyond such questions, smoothing over unexpected answers. Most of us happily come to believe that our creation was in the stars, just as it should have been.

One day children created by cloning will be as indifferent to their origins as we post-1965-born adults are today. So too for their families, bonding, affection, and the life of the family will have long since carried them beyond such questions, smoothing over unexpected problems.

Compare adults created by in vitro fertilization, for whom we now have a twenty-five-year history. As we know, their special origins to them are no big deal: they are just glad to be alive; indeed, many know they are among the most loved children in human history.

Likewise, and to a much greater degree, originating children by safe cloning will spare them any traumas associated with uncertainty about being wanted. With the exception of firstborn sons of childless monarchs of great kingdoms, such children may be the most wanted, most anticipated children in history.

Children created by in vitro fertilization also know they were wanted, but children created by safe cloning have additional assurance because they were wanted not just as generic children but for their general characteristics. Not only will Faith know that Mom and Dad wanted her to exist, but will know that they wanted a girl with a genetic predisposition to strong religious belief. ("Mom and Daddy always wanted a faith-based girl, so the Lord gave them a way to make me. I love them so much and I sure know they love me.")

Safe cloning will be intrinsically good for children in giving them a deep, rich sense that their parents sincerely wanted them, not only as their firstborn child but also as a their firstborn child with musical talent. Wrapped in parental reassurance, such children might be more self-confident and self-assured than normal children, who must (if they think about it) be uncertain about why they were created.

Beneath everyone's fears about designer babies and unrealistic parental expectations lies something that is different and good: the idea that children are wanted with a white-hot intensity. That is a very new, unusual idea to most people (except perhaps those who are familiar with clinics for assisted reproduction). We need time to be comfortable with the idea that children need not "just come" but can be intensely wanted. Even when parents don't get everything they want, we will learn that wanting kids intensely is a good thing.

CREATING HAPPY PEOPLE AND MORE HAPPINESS

A second argument that reproductive cloning could be in the intrinsic good of children needs a little rope, but not too much. That rope is the assumption that human genetics, proteinomics, and knowledge of functional gene-environment interactions will increasingly reveal not only the causes of bad states, such as depression and hypercholesterolemia, but also of good states, such as being good-natured, generally healthy, and long-lived.

Give me that same assumption here and see if I hang myself. But it's not fair to argue with the assumption that we will come to know which set of genes predisposes people to be happy, healthy, and long-lived. So let's say we do know these genes and how to safely choose them in a child originated by cloning.

Then, ceteris paribus, it would be intrinsically good for the child to have genes that predispose her to being happy, healthy, and long-lived rather than genes that predispose her to being morose and sick, and to die early. Such genes indeed may be *foundational* for a happy life, something that over time we come to see as each child's birthright. One day in the future, getting a healthy gene pack will be like getting fluoride in the water to prevent cavities or getting standard vaccinations against deadly childhood diseases.

Moreover, to the extent that children created by cloning are happy, the general happiness in the world increases. To the extent that their happiness makes their parents, friends, siblings, grandparents, spouses, and their own children happy, their happiness spreads in expanding circles, creating more happiness. All of this is intrinsically good.

LAST RESORT TO CREATE A FAMILY

A third kind of argument why reproductive cloning could be an intrinsic good builds on the assumption that creating wanted families is intrinsically good. Make this pro-life, pro-family assumption; if we do so, then safe reproductive cloning will almost certainly be a new kind of tool to use to create such families.

Just as insemination of husband's sperm (AIH), anonymous sperm donation (AID), in vitro fertilization (IVF), surrogacy, and ooycte donation were all initially condemned and have now come to be accepted as useful tools in family making, so reproductive cloning one day will be seen the same way. For the one in eleven American couples who are infertile after two years of trying to conceive, how a baby gets created matters very little compared to the fact that the baby has been created.

Another reason why reproductive cloning may come to be seen as intrinsically good and why it should never be taken off the table and treated as a federal crime is that, as environmental toxins and delayed age of attempting first conception rises, citizens of developed societies may one day soon need better tools to aid them in creating children. Canadian sociologists Louise Vandelac and Marie-Helen Bacon of the University of Quebec–Montreal argue that soaring use of environmental pollutants around the world has dramatically decreased fertility in advanced societies.[3] According to these sociologists, increasing breast cancer and endometriosis, as well as declining animal and human sperm counts and potencies, have been "directly associated with the sharp increase in pesticide use and environmental organochlorine chemicals such as polychlorinated biphenyls (PCBs) and hexachlorobenzene (HCB)."[4]

Don't ban reproductive cloning! If we keep polluting our world, as we seem unable to stop doing, we may all be infertile one day and need an asexual way to reproduce humans. (This gives rise to another linkage argument on the other side: clean up the environment or you'll have to accept cloning!)

Cloning as a tool might prevent divorce. Some marriages fall apart owing to a lack of children. In some cases children bind two adults together, getting them over rough times for the sake of the children. This is true for adopted children also; so it is not necessary that the couple created the children themselves.

For some Orthodox Jews and Muslims, being unable to have children is a real curse, in part because the culture and the traditions, to a great extent, revolve around having a family. In such a culture, finding meaning without a family may be difficult. Less commonly, a particular kind of family is highly desired. For example, some marriages may dissolve because only girls are created and no boys. Although feminists argue that cloning should not support sexist choices, isn't it better to have a family based on consenting sexist choices than a divorced single mom or dad? Besides, personal life needs to be cut some space from pervasive moral criticism.

THE INSURANCE ARGUMENT

Another argument may be called the *insurance argument. Washington Post* columnist Abigail Trafford writes that 70,000 American kids under age twenty-five die every year, leaving their families devastated.[5] Previous generations of Americans had large families and frequently experienced the death of a child. Now the birthrate has fallen to 2.13 children for every woman, and many women only bear one child. If that child dies, a woman may be past her reproductive age. For example, Katherine Gordon of Great Falls, Montana, whose seventeen-year-old daughter Emily was killed in a car accident in 1997, wants to clone Emily's genes to create a child. "I know it wouldn't be Emily—it would be her twin sister," she says.[6]

Yes, Trafford agrees, many people find reproductive cloning repugnant, but many medical procedures are more repugnant. We tolerate them to save our lives or, in some cases, to try to create life. In her own case, her maternal grandmother died at age twenty giving birth to her mother, who was raised by her maternal grandparents in Ohio as a late child. Her aunts and uncles treated her as a baby sister and she blended into the family. So life is full of both disasters and surprises, and if we really value families and children, why should we not keep, as insurance, one tool for re-creating both?

PRESUMED CONSENT

For the next argument for the intrinsic goodness of safe cloning, we should imagine not the forward-looking consent of cloned children (since it's nonsensical to talk of the consent of a being who does not exist about whether he should exist) but the backward-looking endorsement of cloned children. Call this the argument from *presumed consent.*

If cloned children later are told of their unique origins, told about why they were created and why their parents chose a particular genotype, and if such children endorse their method of origination, that would certainly be an argument in favor of this method. If most children created by cloning approve of their unique origins, then we can presume their consent for their origination.

The strongest affirmation of the intrinsic good of reproductive cloning would be if children created by cloning used the same method as adults to create their own children. They would not necessarily or even probably clone their own genotypes, but they might indeed clone someone's. If so, we would have something like rich, deep presumed consent.

JUSTICE TO FUTURE GENERATIONS

Another argument why such cloning is intrinsically good concerns the following kind of idea. Although it is controversial, it may be possible to get rough general agreement about what kinds of genotypes are most desirable. We could connect choices of such genotypes to norms of human nature and how it best functions, as well as to ideals of flourishing in human communities. These need not be based on superficialities, media images, or arbitrary qualities (unless one believes *any* such list is going to be arbitrary).

Again, let us separate questions of good from those of right and assume, for the sake of argument, that we agree on an area of values ("good") about human beings. No matter how skeptical the reader might be about the possibility of such agreement, the following thought should still be instructive: if we could agree about such values, wouldn't it be right for society to encourage their creation? To have more of the desirable genotypes rather than less?

This seems to be the idea of having a convocation of people over generations come together under a veil of ignorance, as in the philosophy of John Rawls.[7] In such a convocation, wouldn't we choose to have children in future generations as healthy, smart, and happy as possible? Don't we owe them this, if we can choose it?

Suppose such genotypes are defined minimally: absence of major genetic diseases, tendency to be happy and long-lived, average to above average intelligence. Again, wouldn't it be not only not bad but actually just for society to create more rather than less of such beings?

Consider what I shall call *the paradox of improvement.* It arises in thinking this way: when it comes to improving the human race, most people would agree with the idea that a thousand years from now, the human race as a whole will be better than its present state. Many might even agree that as soon as a century from now the same human race will be considerably improved.

For example, my fellow biotech optimist, Gregory Stock of UCLA, says, "People will look at this [cloning and germ line genetic enhancements] a thousand centuries from now and will forget all the trauma and the difficulty. The ability to rework our own biology will be one of the basic breakthroughs that laid the foundation of their lives."[8]

But how do we get there from here? When do we start? When it comes to whether Americans or ordinary parents should be entrusted now with improving humanity, everyone says the time is not right; ethics has not caught up with new biotechnology; we should go very slow, use the precautionary principle, and wait until some indefinite time in the future to make any big changes, such as endorsing germ line genetic therapy or human reproductive cloning.

We can see the paradox coming. As once remarked in the cartoon *Pogo,* "We have met the enemy and he is us." How are we supposed to move humanity forward but never be the ones who do the moving? Never be the actual society that chooses to take risks toward moving forward? The paradox is that we want the benefits of an improved humanity without the struggle of getting there, without taking any risks, merely by waving the magic wand of science fiction to transport us there. What magical wisdom are we supposed to acquire in the future that will allow us to know how to change? Why not start now? Maybe wisdom comes in the struggle and is a process, not a static state received from above while meditating on a mountaintop.

Critics can reject the paradox of improvement in two ways. First, they can deny that change can be stopped. This inevitability argument goes the other way, claiming that biotechnological change is coming whether we like it or not, so we had better get ready for it. "Someone will clone a child somewhere in the world, so we had better accept it and move on," one hears frequently.

The inevitability argument is too cynical in viewing human reflection, moral concern, and lawmaking as impotent to stop pernicious biotechnological innovations. If that is so, why am I writing this and why are you reading it? We should just go outside and enjoy the sunshine. A better position is to think about which kinds of change we should embrace, which kinds we should resist, and why.

Critics can attack a second way by rejecting the assumption behind the paradox of improvement—by denying that future humans should improve and differ from us in profound ways. Humans are fine as they are, they might say, as God made them or as they've evolved in nature without God to this point.

This argument can be called the things-are-as-good-as-they-should-get argument, and I suppose it resonates with a reader depending on whether she sees the current glass as half full or half empty. Certainly many things are good for citizens of North America and Europe who live in democracies with good economies. But the rest of the world is another matter.

Certainly biotechnology, which includes genetically modified foods, pharming (growing hard-to-make drugs in special genetically modified fields on special farms), cloning, and animal cloning will be useful tools in raising the standard of living in developing countries. They will also help cure the myriad diseases and dysfunctions (e.g., spinal cord injuries) that currently afflict the lives of many people.

A century ago, critics argued that American society had gone about as far as it should go, and when automobiles, anesthesia, and airplanes were introduced, they were resisted by those who didn't want to change. Will it ever be easier to be a proponent of taking risks rather than a critic of those who want to do so? The easiest thing in the world is to be a critic. Leon Kass, Charles Krauthammer, Francis Fukuyama, Wesley Smith, Bill McKibben are canaries singing the same nay-saying song.

We frequently hear that improvement is inevitable because of a technological juggernaut. That view is false. Artificial insemination of husband's sperm (AIH) was stopped for a hundred years in America, preventing thousands of wanted American babies from being born. People fought against and suppressed latex condoms in the 1930s and birth control pills in the 1960s. For over twenty-five years, physicians have not taught their patients that they can double up on birth control pills and use them as a morning-after pill after a night of unprotected sex. People still wrongly believe that terminal people suffer when taken off artificial food, not knowing that artificial feeding can actually make such people suffer (too bad the parents of Terry Schiavo didn't know this). When it comes to dissemination of knowledge and education, nothing in human progress is inevitable.

But doesn't this move us into eugenics, which is a taboo topic? Like cloning, the "eu" word is a trump card in debate because everyone assumes we can't go there. If you use the word, as Nobel Prize–winner Joshua Lederberg once did, you get labeled as a kook and people imply you have Nazi sympathies. If we are ever going to move humanity forward, we have to transcend the association of making humans better with knee-jerk invocations of fascism and Nazi eugenics. We owe it to future people to make them better, and it is just to do so.

But, one might object, as a matter of justice, doesn't making people with better genes send the wrong message to people with gene-based disabilities? Telling disabled kids that, if given the choice, their parents would choose healthy kids over them? I'm not sure this is as big a problem as I once thought. Doesn't it also "send the wrong message" to allow healthy kids to play soccer but not integrate kids in wheelchairs into the soccer game? To have programs and schools for gifted kids separate from those for normal kids, or special programs for developmentally delayed kids?

For all the above reasons, cloning has come to symbolize more than a novel way of originating a child from the genotype of a previously existing person. It is linked to expanding the range of parental choice, to a possible genetic overclass and normal class, and to some topics we have never found it easy to talk about in North America and Europe: eugenics and why people have kids. Nevertheless, decades after it becomes safe, I predict more and more people will see it as a useful reproductive tool, as just another way to create a family—an intrinsically good thing for medical science to give us.

INSURANCE TO SAVE HUMANITY

It is amusing in a terrifying way that in 1980, specialists in human infectious diseases (ID) thought that all known lethal infections had been discovered and were under control. At the time, ID was considered a boring medical specialty. Then along came AIDS and no one thought ID was boring any more.

By 1987, we were shocked that 60,000 Americans had died of AIDS, more than had died in Vietnam, and researchers thought that 10 million people might be infected worldwide. By 2001, half a million Americans and 20 million people worldwide were dead from AIDS, with another 40 million infected worldwide. If past trends continue, those infected now may pass the virus to another 50–60 million people. In the late 1990s, no one believed a protein could be infectious; then Stanley Pruisner got a Nobel Prize for proving it so. Now we also know about mad cow disease and how prions carry encephalopathies.

In 2003, SARS came roaring out of nowhere to teach ID physicians another lesson in humility. Now we fear it will return each winter. Experts in pandemics predict another lethal influenza virus will sweep the world of the kind that hit in 1918 with the Spanish flu epidemic that sickened an estimated 20–40 percent of the worldwide population and killed 20 million.[9] Other great flu pandemics of the past hundred years occurred in 1957 (Asian flu, killing 70,000 Americans) and 1968 (Hong Kong flu, killing 34,000 Americans). Dr. Greg Poland of the Mayo Clinic says it will happen again in the next decades and be "horrific."

As we will see in chapter 15, xenotransplants may inadvertently allow a lethal infectious virus to jump from nonhuman animals to humans, creating a pandemic. This might be an especially hard virus to contain or stop. What does all this have to do with cloning? The answer is that cloning is a tool that allows new genes to be precisely inserted into embryos. It does not take too much imagination to see that when a lethal viral epidemic, call it "Flu X," starts slaying half the humans on earth, everyone is going to look for something to ensure that humans continue to live, especially humans in their own family, group, and nation.

After the first decade of AIDS, it became apparent that some HIV-infected people were asymptomatic a decade after getting infected. Soon it was learned that their unique genetic composition made them immune from the effects of HIV. When Flu X hits, it is almost certain that some humans with a slight genetic variation will be immune to it, not even getting sick. These are the people from whom human embryos could be cloned to give resulting adults immunity.

Even worse, suppose that Flu X hits newborn babies especially hard, killing 95 percent. At that point, wouldn't we be glad that we had a tool to create new human babies who would be immune from Flu X and live to become adults and who could start to be a new generation to take care of existing adults when they got old? To inherit civilization and our developed countries with all their wealth and technology? Yes, it's an example from science fiction, but AIDS, SARS, and flu pandemics could be the precursor to something much bigger. One thing is for sure: once it hits, it will be too late to do research in cloning. For humanity to have a chance, we need to develop this tool now.

EXPANDING CHOICE

Cloning would expand the currently small range of choices about creating children. It would also demystify the process of making children, taking the religious mumbo jumbo out of creation. Consider an analogy: imagine if the only way you could get a car (and this may have been true at one time in some communist countries) was through a lottery from the central allocation authority. If you were a fatalist, you would accept whatever you got, saying, "It's God's will that I drive a Hyundai."

Present conception is a lot like that. Whether you get a child, and what characteristics it has, are determined by other factors, not the parents. Many people now see that as a good thing and think it bad that parents might soon be able to control the kind of kids they have. But that is because everything is new and people fear that parents wouldn't love children chosen for their traits. But just as people once feared that parents wouldn't love their children if the children were planned, so too people will learn that chosen children will be loved more, not less, than those who came unplanned and unchosen.

Because cloning involves a choice about which genotype to reproduce, it would remove the arbitrariness of genetic roulette in sexual reproduction. It would put our growing knowledge of biology and genetics, not religion, behind choices about children. Although religion may be a force for good once children exist, it has generally been a force for evil in blocking scientific ways to overcome infertility or new ways of creating them. Reproductive fatalism is a flawed worldview, asserting that people should accept disease, dysfunction, and infertility. But this is false; all the above are human evils and medicine's enemies.

Reproductive fatalism holds that the status quo is natural and good. Although life now is much better than it was two or three hundred years ago, we know that life now is as good as it gets. To change any more, especially by adopting radical new kinds of biotechnology, is to risk losing everything.

That view is silly. It is almost certainly true that the present state of humanity can be improved in a thousand important ways. No one deserves to be sick, crippled, or barren. No one desires to die from old age or disease. Reproductive fatalists say we don't know enough, are not wise enough, to know when to change. But change in itself is not bad; we should adopt an experimental attitude toward change: judge each change by its consequences. Changes can be reversed if they work out badly (Prohibition, untaxed cigarettes).

Nor are the most primitive ways of conceiving children the best ("It was good enough for me"). Having lots of ways to create wanted children is good. Some people will always use the most primitive ways; others will study all available options and choose the methods best for them. For still others, need will drive their choice.

Humans are not, as reproductive fatalists subtly imply, basically bad. When most of us think about it, we really don't believe anymore in Augustine's view of original sin, that human nature is tainted by terrible flaws deep within. No, humans are basically neutral to good. Yes, they are self-interested and possess only limited altruism, but that just proves that humans are neither saints nor perfect. Humans do have the moral ability to judge each case affecting them from experience, compassion, and reasoning, and then to blend many values to get the best answer. This is also true about choices about reproduction.

Similarly, reproductive fatalists and liberals do not want to trust parents to make choices about children, especially traits of children. They see all parents as potential child abusers who need to be monitored by the state. But that view is false to most experience. Most parents are good and want the best for their kids. Hence we can trust most parents without state interference or state regulation to make the best decisions about when and how to create children and how to raise them.

CLONING COULD USE ADOPTION

Some bioethicists have proposed that cloning and other reproductive technologies be regulated similarly to adoption—interviewing prospective parents, weeding out inappropriate parents, and supervising the match for the good of the child. Narcissistic parents who wanted a genetic copy of one of themselves, or parents with unreasonable expectations, would not be allowed to use cloning to create children.

This is a possible compromise that would allow reproductive cloning but keep it narrowly confined to great parents who use reproductive cloning as a last resort to have children genetically related to them or perhaps (if this is deemed an acceptable reason) children who will continue an ancient family line. The odd thing about this approach, of course, is that the child does not yet exist. It is like Snowflake, the embryo adoption program championed by conservative Christians, which allows couples to pay for the adoption of an embryo that might, if appropriate gestation were found and if it did not miscarry, one day become a baby.

Although adoption sounds workable, I suspect that it is not. Cloning and adoption are already plagued by the politics of abortion, and cloning is such a sensationalized topic that finding acceptable reasons to use it might be nigh impossible. It would be like convincing hospitals before the 1973 *Roe v. Wade* decision of the need for a therapeutic abortion because of maternal health concerns or a monstrously abnormal baby. Even after taking thalidomide and having a fetus that lacked two legs and one arm, Sherri Finkbine in 1962 couldn't convince a Phoenix hospital to do an abortion and had to go to Sweden to have it aborted.[10]

REPRODUCTIVE CLONING EXPRESSES HUMAN NATURE

One of the leading theories of human nature is that humans are the tool-using animal. From the first primate who picked up a jawbone to use as a weapon to the early man who created a makeshift stool to reach a banana, humans have distinguished themselves by perfecting tools to control their environment, feed their young, and extend their lives. Reproductive cloning clearly fits our essence, for we are compelled by that essence to seek better tools to reproduce ourselves. Not only is reproductive cloning not an affront to human dignity, but casting off this tool would be an affront to it, to not seek new reproductive tools and, most horribly, to destroy new tools recently acquired.

INJUSTICE AND GOOD THINGS

A number of egalitarian critics of cloning primarily object to reproductive cloning not because it is unsafe, damaging to the child in some way or the other, or pernicious to the family. Instead they appeal to various ideas of social justice and argue that such cloning would only benefit the rich. This is a common objection from Marxist professors and social justice theologians. Because cloning would create a permanent biological inequality much worse than mere environmental in equality, they say, it should be banned.

Regardless of the merits of this argument on its own (which I've noted and discussed elsewhere), it is an indirect argument that reproductive cloning would be a good thing. Egalitarians don't want the rich to have this new tool because they don't want rich kids to have more advantages over poor kids than they now have, which of course implies that cloning is a good way for a rich kid to be originated.

Critics of cloning often make the egalitarian criticism and also claim that reproductive cloning is in some way evil. But that is grossly contradictory: if it is an evil thing, then denying it to rich kids is no injustice; if it is a good thing, then it is not evil. Indeed, if it were an evil thing, say, as a likely harm, then originating kids this way would be just the opposite of the usual claim about injustice in medicine. Because they have nowhere else to go, the poor often enroll in public teaching hospitals that simultaneously carry out medical experiments. Champions of social justice have decried this situation, arguing that the harmful burdens of medical experimentation should be shared by all of society and all children.

But if reproductive cloning really is likely to harm children originated this way, then the opposite situation will occur: children of the rich will suffer first, and if such cloning only occurs through private, off-shore in vitro fertilization clinics, then only rich people will be able to avail themselves of such services. But then the claim about social justice must be dropped.

Advocates for social justice could retort that, yes, perhaps that's true in the short run, but when cloning eventually proves successful, as it will if the rich fund it and use it for their own children, then at that point it will become another weapon in class warfare. At that point, we will have a biological innovation that makes society more, not less, unjust, so in anticipation of this endpoint, we oppose it now."

My reply to that retort would involve the scale-to-issue problem mentioned two chapters back. To assume that successful reproductive cloning would only be available to the rich, and not also to middle-class people who valued certain kinds of children highly enough to forgo other goods (like the current situation with in vitro fertilization), is to make large-scale assumptions about human behavior and access to biotechnology that probably will not hold up to scrutiny.

Critics must choose their weapons: either the liberal-egalitarian one that assumes cloning is good but indirectly bad because it furthers inegalitarian injustice, or that it's bad but lacks import for social justice.

CONCLUSION

Once the emotion and sensationalism are stripped from the topic of safe, reproductive human cloning, it is surprisingly easy to justify as something intrinsically good for humans. That is because it would be just another tool in our reproductive tool kit for creating families and better humans. And how can those two things not be good?

12

Risking My Baby

Human dignity in creation requires information, choice, and a judicious balancing of competing traits. Being created through the capricious, arbitrary mix-up of helical chemicals undermines human dignity. How could humans have endured such originations for so many centuries? The simple answer is: they had no other choice.

—United Nations Report on Global Trends, *2150*

Do parents have the right to risk a defective child originated by cloning? Isn't it obvious that they do *not* and that any such attempt should be a crime?

Although this book does not discuss the current state of mammalian cloning and the current hypotheses about whether all cloned mammals are abnormal, it does discuss the politics of this science, and in that vein, we must note that parents desiring to clone a baby are now being asked to jump over a very high bar of evidence indeed, much higher than what we now ask of other parents. Not only are predictions of harm to babies from cloning politicized, and therefore likely exaggerated (especially when implied to be insurmountable), but the process of allowing parents to decide about risk is also politicized, as we shall see.

THE INTUITIVE VIEW

It probably seems obvious to most people that reproductive liberty does not extend to the right to attempt to originate a child by cloning because we currently believe that such a process carries a high risk of producing a defective child. Call this the *intuitive view.*

Supplying a consistent ethical justification for the intuitive view creates problems about consistency. Although some people may be unbothered by lack of consistency in their moral positions, if we are asking for reflection on this issue, such lack of consistency should bother us. Call this the *consistency test.*

Consider an obvious justification for making such actions a crime. It is natural to think that reproductive liberty only encompasses the right to attempt to have a baby if reasonable medical evidence exists that the child will not be born with the kind of gross genetic defect seen in spontaneously aborting fetuses in most cloned mammals: oversize offspring, epigenetic reprogramming errors, and so on. Because good scientific evidence exists now that most children born from cloning would have a genetic defect, or the odds for a particular child are quite high that it would have such a defect, to attempt cloning now is immoral. Hence one could seemingly argue from that conclusion to the even stronger conclusion that such attempts should be criminalized.

Yet that cannot be the reason why reproductive cloning is wrong. At least, it cannot be the reason if we are not guilty of selectively applying a justification to a new reproductive technology with sensational emotional associations and ignoring its obvious applications to a myriad number of ordinary other cases in reproductive ethics. Consider present testing of adults for terrible genetic diseases such as Huntington's disease, medullary thyroid cancer, and achondroplasia (dwarfism). Once we test such adults, we know who among them has a 50 percent chance of passing on their disease to each of their children (Huntington's, achrondoplasia). If reproductive liberty only encompasses the right to attempt to have a baby "if reasonable medical evidence exists that the child will not be born with a gross genetic defect," then it should be a crime for parents carrying genes for these diseases to create children. Chances are 50 percent that their children will be born with a preventable disease.

Such parents could be offered prenatal testing for the diseases and free abortions for any fetus testing positive, such that disease-free fetuses could be carried to term. But the underlying ethical justification entails that if "reasonable medical evidence" exists that the child will enter the world with a serious genetic defect, it is wrong to bring it into the world, or to attempt to do so. From this moral conclusion, we could also move to the stronger conclusion that it should be illegal to do so.

On the other hand, it is not uncommon for parents to approach genetic counselors when pregnant or thinking about pregnancy and ask for testing to rule out normal fetuses. In my medical school during the fall of 2003, we discussed a case in the genetics course of parents with achondroplasia who only wanted a similar child. Like well-publicized cases of deaf parents who only wanted a deaf baby, the achondroplastic parents only wanted a "little person" child. Our genetic counselors did not forbid these parents from seeking children such as themselves and it is certainly not a crime to do so.

Indeed, over the past two decades, one of the great changes in this area of ethics is the rise of disability culture. People with disabilities and their families have become forceful advocates that their condition is not a problem; rather, it is society's attitude and lack of funding that makes their condition a problem. Just as medicine once regarded being gay or lesbian as a disease and eventually gave up such prejudicial attitudes, so disability advocates expect medicine one day to drop its prejudiced views of their conditions. Given such advocates, and given the rise of disability culture, it is more and more common for parents with autosomal dominant diseases to sometimes ask for testing to select babies with their same condition.

In the world of genetic counseling, it is considered unethical to pressure anyone to have an abortion for eugenic reasons. Likewise, many people would find it abhorrent that a woman with an inherited genetic disease could be put in prison for having a baby. The whole enterprise of genetic counseling of couples has adopted the ethical position that, even if couples refuse on principle to abort, they should be given news about the inherited genetic diseases of their fetuses. Such news, it is argued, can help such couples plan for a child with special needs.

Our problem in ethics is consistency, of explaining why we should not punish a couple who refuse to abort a fetus with Huntington's but should punish a couple who try to clone the genes of a healthy adult. Can this asymmetry be explained? A sociologist might suggest that for various reasons we often demand a higher standard for a new reproductive technology (public support for its research, fear of slippery slopes, etc.) than for a traditional way of conceiving babies. But that explanation is not an ethical justification; it is just an observation about our inherited fears, conditionings, and feelings.

Certainly, if we adopt the ethical criterion of what is in the best interests of the child-to-be, then it is hard to see why the onus of proof is on the new reproductive technology. In the case of Huntington's, the fetus that tests positive is virtually certain to have a terrible disease, whereas this is not true of a cloned human fetus that does not miscarry and makes it to birth.

Every day, thousands of babies are born harmed because their mothers smoked, drank alcohol, and/or used damaging drugs during pregnancy. These mothers knew that "reasonable medical evidence" indicated that their behavior was harmful to their baby, yet they did not change. If our justification for not allowing parents to have a cloned baby is the above, how can we not make it a crime to smoke, drink alcohol, or get high on cocaine while pregnant? Aren't we being inconsistent and hypocritical in criminalizing cloning? Even if mothers in the past didn't know how damaging their actions were, we know now. Alcohol is probably the most toxic drug to a developing fetus, even in as little as a drink and a half a week by the gestating mother.[1]

Furthermore, it's not that society never indicts parents for ignoring reasonable medical evidence. Consider the Christian Scientists who were indicted and convicted by a jury of manslaughter when they refused to take their child with a high temperature to a hospital (the child had meningitis and died when they took him to a faith healer). In another case, a father was convicted who refused chemotherapy for his child with leukemia (a disease that in children is curable by chemotherapy). We also overrule parents who are Jehovah's Witnesses who want to let their child die rather than subject the child to a blood transfusion, reasoning that the child as an adult might not choose to be a Witness. Mormon fundamentalists are prosecuted who will not take their children to hospitals for medical treatment for emergencies. In all these cases, we reason that preserving a child's life—where reasonable medical evidence indicates it likely will be preserved—should overrule religious liberty.

But of course there is more going here. The three religions mentioned above are considered cults by a majority of Americans. Cloning is similarly perceived as something only Raelians or extreme narcissists would want. The rule here seems to be that reproductive liberty of prospective parents can be squashed if we are dealing with eccentric parents in a cult.

But what kind of moral principle asserts that if you are parents in a minority religion, or want to try to originate babies in a new way and in a minority, your reproductive liberty doesn't exist? It is not that we allow reproductive liberty to be overruled because we want to prevent harm to the child-to-be, because otherwise we would criminalize dysgenic births for ingesting harmful substances during pregnancy. No, we pick on new reproductive technologies and people in minorities, making them jump over a higher bar or just forbidding them from having babies under threat of jail.

All that is quite inconsistent. So the problem remains that we do not know how to ban reproductive cloning without banning a huge range of present behaviors that we have long accepted in mothers and couples but is equally, if not more, harmful to children.

WHO DECIDES HOW MUCH RISK TO TAKE?

A quite different but related question is who decides how much risk is ethically and legally permissible for parents to take in conceiving or raising children. The advantage of classifying the questions this way is that it highlights the fact that we require stronger justification for making something illegal than for considering it unethical. If you have the means to do so, it may be unethical for you not to financially support your children in college but it is not like failing to pay court-ordered child support. It is unethical never to take your minor child to a physician but it is not illegal. And so on for many other activities of parents and children.

In a similar manner, once children exist, we have stronger standards about harm than when they are fetuses or embryos, and we are even less strict about possible harms involving conception where without those harms, the children would not exist at all. So it is a crime not to provide some sort of schooling for children, but it is not a crime to forgo taking folic acid during pregnancy to avoid neural-tube defects such as spina bifida in newborns.

Let's now return to reproductive cloning. Even though this involves a child-to-be who would not exist without the novel method of his conception, we inconsistently require a higher moral standard for allowing it than for nontreatment of existing children or for failure to benefit existing children.

We have skipped over a huge area about who decides how safe reproductive cloning must be before parents are allowed to attempt it, whether we take norms from teenagers in sexual reproduction or forty-year-old couples who are mostly infertile (with much higher rates of complications). And given the politicization of this whole debate and its links with abortion, what chance is there of arriving at a reasonable standard?

So we are back to the problem with which we began, namely, why do we require Olympian moral standards for allowing parents to risk conceiving a child through cloning when we tolerate great parental neglect and harm with sexual reproduction/gestation and with existing children? Although "two wrongs do not make a right," surely we require some consistency in reproductive ethics. If we are wildly inconsistent about cloning, might it be because its sensationalized history and publicity blind us to seeing it as something not too different from other things we now do?

SCALE-TO-ISSUE PROBLEMS, AGAIN

Once again, we see public policy distorted by a sensationalistic case while the banal reality of thousands of others is ignored. While the whole planet seems focused on preventing one human cloned baby, which may or may not be defective, it ignores the thousands of babies born daily who are hurt by parents who smoke, drink, or take teratogenic drugs.

In another whole class of harms, we will one day know how to screen embryos with gene chips for hundreds or maybe thousands of genetic diseases, making another kind of harm preventable. But it is unlikely that deciding to implant a "diseased embryo" will be made criminal.

So thousands, maybe millions, of babies will continue to be born as we fail to reexamine our traditional intuitions about why and how babies are born. Way too much attention has been spent on cloning—on preventing at most a few dozen human babies from being born who, if really abnormal, will secure the wrath of the world on the parents and scientists who produce them. And that will be that.

If the intuitive view is correct, then a consistent ethics should either allow parents to originate children by cloning or ban parents from gestating an embryo with a known genetic disease. If originating a child by cloning is wrong because of the likelihood of defect, then a much larger class of human gestation is also wrong. Put differently, when it comes to thinking about the ethics of how babies are created, most of us fail the consistency test.

13

Should Cloned Children Be Outlaws?

The law exists as a kind of outer limit on morality: certain kinds of offense and slight count as unethical but not as illegal. When it comes to gross harms, the law steps in. Its chief mechanisms work well in so far as life proceeds normally with ordinary thieves, batterers, and scoundrels, whom lawyers, legislators, and judges understand all too well. It completely breaks down with cutting-edge breakthroughs in the bio-sciences, which befuddle almost everyone in American law, politicizing the process to the detriment of humanity, and shackling elite researchers in religious leg-irons.

—*Judge Learned Foote, Dissenting Opinion 2059*

Most people assume the federal government has the right to make cloning people illegal. They imagine that government can control mad scientists growing people in cavernous labs and they see doing so as a legitimate power of government. After all, government exists in part to ward off terrible horrors and avoid having us all return to a state of nature.

Nevertheless, the federal government may not in fact have the right to criminalize reproductive cloning. The U.S. Constitution, Bill of Rights, and over two hundred years' worth of decisions by the U.S. Supreme Court do not clearly enunciate the power to do so.

Even if it were ambiguous, it is wrong that the government should exercise its power to make cloning a federal crime. Is it really the best long-term public policy for the federal government to criminalize any area of human reproduction? I strongly believe it is not, first because use of this cumbersome federal club is unnecessary and, like the arms race, once started, is hard to stop. Second, it could create a set of disenfranchised human beings who lack the common legal rights of all other Americans (more on this later).

Of course, states may regulate the behavior of physicians: they may require them to be licensed, not do certain kinds of abortions, register to prescribe narcotics, and so on. States may regulate or ban ways of originating children, and to date six states, including California in 2002, have criminalized reproductive cloning.

The government may also regulate medical experiments in institutions receiving federal funds through the Food and Drug Administration (FDA) and Health and Human Services Department (HHS). The FDA, which has authority over the introduction of drugs and new devices, claims that it has the authority to regulate reproductive cloning and, in particular, require an investigational new drug (IND) application for anyone wishing to clone a human baby. For this reason, the FDA asserts, the Raelians cannot clone a baby in the United States (since the FDA assertion, Raelians have not claimed to do so in the United States—not that their claims are real).

But the authority of the FDA over groups *not* receiving federal funds is unclear, and if every state did not criminalize reproductive cloning, then the lack of a federal prohibition would be important. As I will argue soon, such a law would be unconstitutional. Before that, I want to emphasize one last point, which is suggested in the title of this chapter. Harvard law professor Laurence Tribe emphasizes that bans on cloning

> will not be airtight. Just as was true of bans on abortion and on sex outside marriage, bans on human cloning are bound to be hard to enforce. And that in turn requires us to think in terms of potential outcasts—people whose very existence society will have chosen to label as a misfortune and, in essence, to condemn.[1]

In this chapter I argue that, given all the existing remedies for dealing with reproductive cloning, given all the overblown hype about such cloning that will eventually die down, we should not burden children created by cloning with what Professor Tribe called "outcast" status. As Yale professor Nick Bostrom argues, "How would you feel [as a human originated in the future by cloning] to hear all these dignified people talking about you as if your very existence were a crime against humanity?"[2] Professor Bostrom then argues that the dignity of humans requires not making cloned beings illegal.

Condoning the creation of a class of persons with legally dubious origins would be a step backward. Over the centuries, Western societies have evolved a very important principle that has slowly evolved in mature ethics: that how one is originated as a person—whether one's parents were married, of the same sex, of the same race, spoke English, were legal immigrants—does not affect one's moral status. If one is born of a human, one is a moral human with full rights.

In a later chapter, I will argue that we should at present extend this principle now to human chimeras. Rather than dehumanize such beings by labeling them as outcasts, we should lift them up in a preemptive move to full human moral status. Taking a step back from that principle is no small matter, and certainly the reasons for criminalizing reproductive cloning do not justify it.

THE HUMAN CLONING PROHIBITION ACT OF 2001

The Human Cloning Prohibition Act of 2001, reintroduced as the Human Cloning Prohibition Act of 2003 (H.R. 534 and 801), states: "It shall be unlawful for any person or entity, public or private, in or affecting interstate commerce, knowingly (1) to perform or attempt to perform human cloning; (2) to participate in an attempt to perform human cloning; or (3) to ship or receive for any purpose an embryo produced by human cloning or any product derived from such embryo." It also states, "It shall be unlawful for any person or entity, public or private, knowingly to import for any purpose an embryo produced by human cloning, or any product derived from such embryo." This act was passed in the House of Representatives but has never been passed in the Senate.

Violation of this law carries not just civil but criminal penalties: "Any person or entity that violates this section shall be fined under this title or imprisoned not more than 10 years, or both." On the civil side, "Any person or entity that violates any provision of this section shall be subject to, in the case of a violation that involves the derivation of a pecuniary gain, a civil penalty of not less than $1,000,000 and not more than an amount equal to the amount of the gross gain multiplied by 2, if that amount is greater than $1,000,000." So a physician, citizen, or researcher can be sent to jail for ten years or fined $1,000,000 for trying to create a cloned embryo, gestating one, or bringing one into the United States from abroad. If a woman goes to England, where embryonic cloning is legal, has cloned embryos implanted in her uterus, and returns to the United States, immigration inspection at an international airport might be interesting.

JUSTIFICATIONS FOR OUTLAWING CLONING

Now that we have seen what the Human Cloning Prohibition Act outlaws and with what penalties, we need to look at its shaky justification and whether such justification is consistent with other interpretations of the constitutional powers of Congress.

Intervention by government into citizens' private lives requires strong justification. We are a free country of free citizens who cherish our hard-won liberties. Every time governments take away liberties, or try to, it is always for a good moral purpose. But the ends rarely justify the means.

In a constitutional democracy dedicated neither to promoting tacitly understood theistic values nor to expanding federal power, protection of individual life and liberty are arguably the two most sacred purposes of the rule of law. The liberty to decide whether, when, and how to have children is perhaps our most important liberty, other than the liberty to life itself.

Several law professors argue that such reproductive liberty amounts to a fundamental interest under the due process clause of the Fifth and Fourteenth Amendments. These professors include John Robertson of the University of Texas Law School, Elizabeth Price Foley of Michigan State University, David Orentlicher of the Indiana University School of Law, Robert Moffat of the University of Florida College of Law, John Charles Kunich of the Roger Williams University, and California trial lawyer Mark Eibert (an advocate for infertile people).[3]

According to traditional reasoning in decisions by the U.S. Supreme Court, bodies such as Congress may only intrude on a fundamental interest by passing the standard of strict scrutiny. This means that government must show *clear and convincing* or *compelling* evidence (italicized words have technical legal meaning) why the intrusion is warranted, and possibly evidence *beyond a reasonable doubt* that such intrusion must be allowed.

More technically, in a substantive or major case affecting due process, for the government to overrule a fundamental constitutional right, it must show that its law or action is (1) necessary to a compelling state interest and (2) narrowly tailored to achieve that specific end. All these carefully constructed restrictions create a high hurdle to jump when the federal government tries to curtail personal liberties affecting reproduction.

Certainly some past decisions by the U.S. Supreme Court can be interpreted this way. Suffolk University law professor Barry Brown observes that in *Lifchez v. Hartigan,* a U.S. District Court emphasized this interpretation of the right to privacy: "It takes no great leap of logic to see that within the cluster of constitutionally protected choices that includes the right to have access to contraceptives, there must be included within that cluster the right to submit to a medical procedure that may bring about, rather than prevent, pregnancy."[4] More powerfully, and as California lawyer Mark Eibert emphasizes, the U.S. Supreme Court argued similarly in *Eisenstadt v. Baird* that "if the right of privacy means anything, it is the right of the individual, married or single, to be free from unwarranted governmental intrusion into matters so profoundly affecting a person as the decision whether to bear or beget a child."[5] For Eibert, *Lifchez, Eisenstadt,* and a decision preventing Oklahoma from sterilizing a prisoner for eugenic reasons because it violated his fundamental "right to have offspring,"[6] taken together, give Americans the "right to a biologically related family." If Eibert is correct about this claim, then couples have a right to attempt cloning as part of their basic liberties as Americans.

Even if cloning does not fall under the protected fundamental liberties of Americans, it may fall under the broader kind of liberty that defines the pursuit of happiness. Government may still ban cloning, but here the bar is lower than if such a ban competes against a fundamental liberty, it must go. Even a ban under a lower standard of evidence would need to show some great harm that could not be prevented by other means (e.g., malpractice suits; see below), such that federal government must intrude into citizens' personal lives.

Much federal government expansion has been justified in the past by invoking the commerce clause of the Constitution, which allows regulation of interstate and foreign commerce. If the federal government had the authority to regulate cloning or reproductive clinics, that de jure power would have to come from somewhere and critics of cloning would likely point to the commerce clause. In recent years, the U.S. Supreme Court has been much less sympathetic to such arguments.

INTERLUDE: WE NEED LIBERTARIAN REPRODUCTIVE RIGHTS

What right does government have to enter your home and tell you how to have kids or not have kids? Suppose the process of creating a cloned embryo becomes as simple as using a turkey baster, which some women use to perform artificial insemination of sperm (a kid has already cloned a cow, showing that it doesn't take a genius to clone animals). How can government enforce a ban on cloning without invading privacy and trampling personal liberty and turning American democracy into a police state? Is preventing cloning worth that?

Once you accept the idea that government has a moral and legal right to tell you how to conceive, you are doomed. To see why, suppose that reproductive cloning becomes as easy and as safe as present human sexual reproduction. If the federal government can regulate asexual reproduction in the private home or lab, then why doesn't the same principle give it the right to put something in the water to regulate sexual reproduction and decide who gets to reproduce and who does not? I don't see how the principle can be interpreted only to ban asexual cloning without also threatening sexual reproduction.

Indeed, if government has the power to prevent future babies from being harmed, why doesn't it also have the power to ensure that future babies are benefited? To come into your house and impregnate you with the approved sperm? Why do the *location* of the citizen and the *method* of conception of the embryo matter? Why should those two factors carry so much moral weight and even more legal weight? If government can criminalize conception of one kind of embryo in a reproductive clinic, why can't it criminalize creation of another kind of embryo in your home? Does the location of the conception matter to the harm done to the child? If a physician in reproductive medicine puts a cloned embryo into her uterus and gestates it, has she done something different than could be done in a clinic for assisted reproduction?

BACK TO THE CONSTITUTION

Does clear and convincing evidence exist that parents who would clone children are so deranged that they should be made into federal criminals? No. Even if originating children is at present almost always harmful to children, the federal government still has no right to tell physicians how to practice medicine or to tell parents how to originate children. Perhaps raising children as atheists or home-schooling them harms them, but government should not interfere.

More to the point and as argued in the previous chapter, any couple who uses presymptomatic genetic diagnosis for an inherited genetic disease and chooses not to abort is creating children burdened with a harm deliberately accepted.

We have called this the *consistency test.* Should government intervene here in family life too? Making cloned children outlaws responds to the emotional punch of "clones" and "cloning." Ignoring the parallel harms in the other case responds to the emotional punch of "eugenics." In either case, the decision is emotional, not rational, and it is certainly irrational not to have a common standard for both cases. Harm to children is harm to children.

One way for the Supreme Court to outlaw cloning but allow medical technology to assist infertile couples is to assert that the fundamental interest in question applies only to sexual reproduction, not the asexual reproduction of cloning. That reasoning, however, spectacularly begs the question. Exactly the same reason could be applied to unassisted versus assisted reproduction: government can ban assisted reproduction but not unassisted reproduction. Doing so puts the weight of evidence on the distinction between assisted and unassisted reproduction; if that distinction won't bear the weight, then everything crashes.

The Supreme Court used the same flimsy approach in its pusillanimous death-and-dying decision in 2001 *(Washington v. Glucksberg)* when it asserted that the penumbra of personal liberty of Americans was not so broad as to allow Americans a right to physician-assisted dying, such that state laws forbidding physicians to do this were constitutional. The court's cowardice came in its justification that such a right not only was absent in American history, law, and ethics, but actually contradicted by the above.

While that is true, it is also true that *every great change in medical ethics, law, or public policy would be forbidden by the same reasoning,* including racial integration, recognition of rights of gays and people with disabilities, concern for animals, and protection of property rights in new technology. This flimsy defense can block virtually any change that comes down the pike and, as such, is a recipe for a stagnant society. Consistent legal decisions based on this reasoning would create a rigid, backward-looking society like that created by Orthodox Islam.

Texas law professor John Robertson offers a compromise, which the high court may one day endorse when cloning primates becomes safe and routine: interpret the reproductive liberty of Americans to include cloning only to the extent that it was the option of last resort to overcome reproductive failure.[7] As such, it will fall under what Eibert called the "right to biologically related children."

The problem with this compromise is that it gets legislature and courts into the business of deciding which reasons are good and which are insufficient for using a human reproduction tool. Perhaps the best way to put the compromise is that even if some risks to the child existed, then infertile parents at last resort, who only want a biologically related child, have the right to try origination of the child by cloning, where they stand ready to assume lifelong responsibility for any mishaps that occur.

EXISTING LEGAL TOOLS CAN DEAL WITH CLONING

We do not need a federal law criminalizing reproductive cloning. The Brownback bill is overkill, fueled by a religious right that, according to Catholic pro-life activist Richard Doerflinger, has made cloning "the pro-life issue of the 21st century."[8] The Human Cloning Prohibition Act is not needed for at least five reasons.

First, and perhaps most important in the lives and practices of actual physicians, existing malpractice law punishes any physician who helps a couple create an impaired child, and because costs of lifelong medical treatment are high in such cases, such judgments are exorbitant. One hears sensationalized charges among physicians even now that attempts to clone a human child would be "child abuse." Given the foreseen risks now of human reproductive cloning, it would be an open-and-shut case of malpractice against a physician who went ahead and helped a couple create a child by cloning, which turned out to be deformed.

Even if the child appeared physically normal, cloning is such an emotional and symbolic topic to many people that a malpractice suit might still go forward on two counts: a lawyer for the child or parents could argue that, even if the baby appeared normal at birth, a physical deformity might show up in the future, and a fund would need to be set up for care of this person when he later became disabled; second, the lawyer could argue that being originated by cloning—as a "copy" of his ancestor—is a psychological assault on the child, who is entitled to damages.

Second, the Food and Drug Administration (FDA) and Office of Protection from Research Risks (OPRR) have publicly argued that attempting to create a child by cloning is attempting an experiment on a child that comes under federal aegis and control, and hence needs the permission of both agencies to proceed. Because of the assertions of these agencies, as well as their threats to prosecute those who ignore them, Zavos and the Raelians no longer claim to have done anything in the United States (although they have actually done nothing, even to claim they had might land them in court to defend themselves against their own claim—as their lawyers surely told them.)

The government regulates medical experiments in institutions receiving federal funds through these two agencies. The FDA, which has authority over the introduction of drugs and new devices, claims that it has the authority to regulate reproductive cloning and in particular to require an investigational new drug (IND) application for anyone wishing to clone a human baby. Even with such a license, approval by institutional review boards (IRBs) and other kinds of ethics committees (e.g., the one of the American Society for Reproductive Medicine) would need to approve the first attempt. (The FDA's claim is very controversial—some would say absurd—because its charter, giving it the right to regulate "drugs and devices," obviously does not extend to regulating experiments on humans.)

Third, dominant images from science fiction of cloned children growing in vats cause people to forget that reproductive cloning requires in vitro fertilization (IVF), that IVF is only successful 20–25 percent of the time, that success requires careful attention from a team of professionals at a clinic for assisted reproduction, and that such clinics screen their customers. The public has a false view of such clinics as producing babies on demand for any nut who walks in with a fistful of cash. That is far from the truth.

Indeed, it is an affront to some Americans that such clinics do not assist just anyone. Almost all clinics require prospective parents to undergo extensive counseling to achieve agreement in advance over number of embryos implanted, what will be done with leftover embryos, and how many times it's reasonable to attempt IVF. Moreover, some clinics do not provide their services to unmarried couples, gay or lesbian couples, crazy people, single parents (because they accept no federal funds—otherwise they could not experiment on human embryos—they are also immune from federal antidiscrimination laws requiring medical services to be provided without discrimination based on marital status or sexual orientation).

In November 2003 a federal court upheld a Denver fertility clinic's decision to refuse to help a thirty-three-year-old blind black woman conceive.[9] An attorney for the clinic said physicians stopped infertility treatments in 1999 after four rounds when Kijuana Chambers wouldn't allow an occupational therapist in her home to evaluate its safety for a future child. Chambers sued in federal court, claiming discrimination under the Americans with Disabilities Act. Physicians at the clinic said turning her down "was the right thing to do" and that the case was about "the moral and ethical responsibility of a physician." Personally, I don't believe this was the correct decision. Do you have to be sighted and partnered in America to have the right to pay to conceive a child? As it turned out, a clinic in Iowa didn't think so and Chambers had a child there two years later.[10]

Hence the bottleneck of IVF clinics is sufficient to control reproductive cloning in North America. Because all such clinics have refused to attempt reproductive cloning, and because any physician who dared to try would bring down the wrath of the FDA and OPRR and malpractice suits, there is no danger that reproductive cloning will take place.

At present, such clinics are not regulated by the federal government and want things to stay that way. Now that Leon Kass has pushed his bioethics commission to seek federal control of such clinics, the pressure is on to try nothing controversial.[11] No clinic wants to rock the boat nationally by attempting to clone a human. Attempting to clone humans would be the surest way to bring down federal regulation of all sorts.

Fourth, medicine does not take place in a vacuum. It is a deeply human enterprise, composed almost entirely of one set of humans providing intensely personal services to other human beings. In this context, we should not ignore or forget the personal ethics of physicians engaged in reproductive work. Most physicians practice ethically and want to do the best for their patients. No physician wants to bring a disabled child into the world. In addition, most physicians are ethically conservative and do not favor human cloning.

Given that a federal court upheld the Denver clinic, and given the political heat from the President's Council on Bioethics to put federal controls on fertility clinics, no hospital's executive committee is going to allow one of its clinics, even a satellite clinic with a loose association, to try to clone a human child. Almost all physicians at present oppose reproductive cloning and would severely censure any physician who attempted it. As with heart transplants, medicine can regulate itself without threat of federal prison.

Fifth, despite the Human Cloning Prohibition Act, states have passed bills regulating or outlawing any kind of cloning, including California, which instituted civil (not criminal) penalties for attempting reproductive cloning. (Whether it's constitutional to ban cloning is another question, but the fact remains that states have done so.) Certainly states can and should regulate the actions of physicians. We don't want quacks delivering our babies or unethical physicians selling notions out of car trunks in front of health food stores.

Given that states already regulate physicians and have established physician boards to hear charges of unethical conduct, they already have adequate legal mechanisms for dealing with eccentric physicians who would attempt reproductive cloning before we have good evidence from primate cloning that it has a good chance of being safe. States presently discipline physicians for unethical behavior, establish laws about malpractice, and distinguish between civil and criminal crimes. Why do we need to go overboard and make cloning a federal crime?

CONCLUSION: MAKE CLONED CHILDREN OUTLAWS?

Given that almost no physician wants to create a damaged child, that no clinic is going to support human cloning, that any physician who did would likely face charges of malpractice or child abuse, and that states are free to outlaw human cloning, why do we need a federal law to this end? The answer is that we do not and it's a mistake to do so.

In general, our law and ethical tradition allow parents to decide how much risk to run in having children with known risks of harm. But we make an exception when it comes to reproductive cloning. That is a mistake. Doing so will ultimately hurt science, will hurt children and adults originated by cloning, is not warranted by the evidence and, in short, at this time is very premature.

14

Prelude to Cloning: The Ethics of Out-of-Body Gestation

And when I was born, I drew in the common air, and fell upon the earth, which is of like nature, and the first voice which I uttered was crying, as all others do. . . .

—*Wisdom of Solomon 7:3*

Creating human beings in entirely new ways creates new ethical issues, or does it? Maybe what seems new is just a function of our ignorance. In thinking about how novel human originations are justified, we should step back and ask: how is any such experiment justified?

During the past century, hundreds, probably thousands, of medical experiments have been performed on human embryos, gestating fetuses, and neonates. Otherwise, no progress would occur in obstetrics and neonatal care. Understanding how research might be justified on out-of-body gestation ("artificial wombs") may discover similarities with the ethics of cloning. To many, the artificial womb is the reductio ad absurdum of medical technology and the opposite of natural motherhood. If research on it can be justified to these people, then perhaps research on human reproductive cloning can be too.

A touchy subject, research on cute, innocent, newborn babies arouses strong protective responses in us. Like human cloning, the artificial womb has a historical pedigree shaping its symbolic connotations. Science fiction has conditioned us to be upset and alarmed about artificial wombs. Indeed, cloning and the artificial womb can with some justification claim to be the primordial cautionary tale of science fiction.

Nearly a century ago in 1924, J. B. S. Haldane, the great British geneticist, wrote *Daedalus*, describing how humanity loved inventors of machines but hated biological innovators such as Daedalus, who oversaw the coupling of Pasiphaë and the Cretan bull to make the Minotaur.[1] Haldane anticipated the "yuck" reaction to everything from in vitro fertilization to cloning and humans with xenografts. Haldane's work is now unknown but his ideas live on through the work they inspired, Aldous Huxley's *Brave New World.*

Although Huxley is sometimes credited with imagining the horrors of standardized out-of-body gestation, Haldane's fictional work, which has the form of a wise narrator looking back from the future to describe key past events in the success of biology, depicts the birth of the first "ectogenetic" child in 1951:

> The problem of the nutrition and support of the embryos was more difficult, and was only solved in the fourth year. Now that the technique is fully developed, we can take an ovary from a woman and keep it growing in a suitable fluid for as long as twenty years, producing a fresh ovum each month, of which 90 per cent can be fertilized, and the embryos grown successfully for nine months, and then brought out into the air.[2]

One thing Haldane certainly got right: the continuing, outsize horror at biological innovations. In our time, people shrug off new machines but grow agitated when someone adds or subtracts a few genes to an embryo in a Petri dish.

Because of such associations, negative feelings about cloning and artificial wombs seem inevitable for the foreseeable future. To many, babies gestated inside cold, inhumane machines, and lacking a mother's warmth and personal concern, is a *Bladerunner* future. For many people, such news would be an emotional kick-in-the-stomach, a frightening threat to their sense of normalcy. To critics, babies raised in such alien environments would necessarily be traumatized.

Like reproductive cloning, the artificial womb is viewed by many as inherently evil, and not a fit subject for research. For them it is intrinsically wrong to even start doing research to develop such a device. As Scott Rae, a fellow at the Center for Bioethics and Human Dignity and a professor of biblical studies, says, "Even if formidable technological obstacles could be overcome, should we move in the direction of developing artificial wombs? Surely this is an example of something that should not be done even if it can."[3] Like research on reproductive cloning, it is an example of what some would call "forbidden knowledge."[4] So powerful is the evil symbolism of the artificial womb that, to many critics, rather than a human good, artificial wombs would be a worst-case scenario.

In contrast, natural motherhood symbolizes what is comforting, safe, and primordial, the opposite of what is dangerous, foreign, and artificial. If ontogeny repeats phylogeny in ethics, our most ancient ethics will be about our primordial roles, as son, parent, teacher, and in our work. For this reason, changes in gestation may seem especially dangerous to conventional notions of family and female nurturing. In this sense, extracorporeal gestation attacks a primordial conception about what we humans are and our world. If I see a young woman with two small children walking down the street, my first thought is of her as the woman who helped create and gestated those children. That may be incorrect, but odds are that it is true.

Extracorporeal gestation is one of many steps away from the primordial sexual picture of reproduction. If concentrating sperm to maximize chances of pregnancy is one step, and in vitro fertilization another, then a continuum exists of assisted reproduction through surrogates and egg donation, the final extreme being gestation outside the female's body. Each step down this continuum introduces more parties (anonymous or designated sperm donors), mishaps (switched embryos), but also babies for more and more people condemned by nature to be childless.

HOW IT WOULD WORK

As already noted, the artificial womb is the ultimate form of assisted reproduction. For over thirty years, medical researchers have been trying to develop one, a feat technically called *ectogenesis.* Some recent research shows promise in this area by creating scaffolds of biodegradable material molded into the shape of the interior of a human uterus, to which embryos (left over from in vitro fertilization) were added.[5]

When Professor Hung-Ching Liu of Cornell University's Center for Reproductive Medicine and Fertility nurtured human embryos with hormones and nutrients, they attached to the artificial uterine wall and grew for six days, until the experiment was halted.[6] Professor Yoshinori Kuwabara was doing similar research in Japan with goat fetuses in artificial tanks filled with amniotic fluids until his death in 2002.[7] Britain's Thomas Shaffer in 1989 tried liquid ventilation on three premature fetuses around twenty-three weeks old, improving their lung expansion, and in 1996 gave the same to thirteen premature infants aged twenty-four to thirty-four weeks; all were expected to die, but amazingly seven lived.[8] Should this kind of research find success, premature babies might be viable who now are not. In addition, as the artificial womb is perfected, it might become an alternative to traditional gestation.

Currently ECMO (extracorporeal membrane oxygenation) functions as an artificial lung for premature babies with inadequate lung development. In function, it resembles the bypass machines used in heart transplantation. After a large tube is inserted in the baby's neck, blood flows by gravity through a plastic bladder, where a membrane oxygenator removes carbon dioxide. Then the oxygenated blood is warmed and returned to the body through an artery in the neck. This machine is now common in all neonatal intensive care units (NICUs) and shows that ectogenesis may not be as far off as commonly thought.

ARTIFICIAL WOMB: BEST INTERESTS OF SOME CHILDREN?

Knee-jerk revulsion to the artificial womb creates the thought that putting a human fetus in one would not be in the best interests of the child. As Christian bioethicist Scott Rae puts it, "Though it may be that such children would not be physically harmed, nurturing them in an artificial womb is far from ideal and is not something we should encourage. Children in the womb are owed the best chance a good start in life, consistent with their dignity as persons made in God's image."[9]

Could ectogenesis ever be in the best interests of the child? Wouldn't natural motherhood always be better? Not necessarily. Women with damaged or dysfunctional wombs will not be able to gestate children. Ditto women with no womb.

Couldn't we use surrogates for such at-risk fetuses? Not often, because women willing to be surrogates are scarce, expensive, and not available on demand when a pregnant woman presents in crisis at the emergency room. Moreover, most mothers of at-risk babies will not be able to hire a surrogate. Besides, surrogacy is not very efficient. So for some embryos, ectogenesis may be the only way they get gestated, and for some women, the only way an embryo from their egg will become a child.

We need to shed the emotion around this topic and simply see extra-corporeal gestation for what it is—a tool to help impaired babies live. Actually, the artificial womb is the ultimate pro-life technology. Indeed, what could be more pro-life? It is, as we shall see below, the opposite of abortion. This (perhaps unexpected) implication shows why no new medical tool should ever be banned, especially those that enable us to start and preserve nascent human life. It seems amazing and contradictory that pro-life bioethicists would ever oppose development of such a technology.

During pregnancy, some mothers unfortunately use alcohol, cocaine, tobacco, and other substances likely to harm their fetus. Because the mother's biological system is the same as the fetus's, whatever risks the mother takes are also borne by the fetus. But the fetus is at a much more vulnerable stage, so the risks of harm from substances are often greater. This context shows easily how artificial wombs could benefit some fetuses. Raised in a uniform, stable, drug-free, controllable environment, they can be spared risks associated with the mother using drugs. For some babies, these risks can be considerable. Women who contract HIV may pass it to their fetuses, and AZT, if used to block transmission of HIV, may harm the fetus.

Of course, the artificial womb itself has risks. It could lose its electrical power, suffer a leak of fluids, or be mismanaged in any of a dozen ways. The question for consideration here will never be whether there is no risk but always whether the risks to the fetus outweigh the benefits. The same control that is possible that allows prevention of toxins entering the fetus's blood also allows for careful monitoring and study of the best possible nutrients for the fetus. We might be able to change the gestational course for a baby of an alcoholic mother from being born addicted to alcohol and retarded to being born alcohol-free and with superior nutrition and oxygenation.

Yet another medical benefit of the artificial womb is that it would allow surgery to correct defects such as hydrocephaly (buildup of fluid inside the brain) or cleft palate. A surgeon could fix a defect in a spinal cord or heart valve much more easily than if she had to cut open a woman's uterus. Such in utero surgery is now done for some fetal defects, but it is dangerous and controversial. It would be much safer if it occurred inside an artificial womb.

Ectogenesis might also benefit the mother. One compelling reason for using the artificial womb is where pregnancy conflicts with the health of the mother or where it might actually hurt or kill the mother. Various medical conditions are exacerbated by nine months of pregnancy, rendering a women's health worse than it was before pregnancy.

Finally, artificial wombs remove all age limits to gestation. In some ways, that might be their most radical implication. After her only existing child was unpredictably killed, a sixty-year-old woman might be able to have her previously frozen embryo gestated in such a womb.

THE ELIMINATION OF ABORTION?

More controversially, a young female who was the victim of rape or incest, who did not want to gestate a fetus, could be allowed to have the fetus transferred to an artificial womb to be given up for adoption. This creates a fascinating prospect. Assume for the moment that artificial wombs are safe and assume that enough funds are available to create one for any woman who wants to use it. Given those assumptions, then the widespread use of artificial wombs might significantly decrease the number of abortions because any woman pregnant against her wishes who did not want to abort could transfer her pregnancy to an artificial womb. Then the child could be adopted by any of the one in twelve childless couples in America today.

Of course this scenario makes a lot of controversial assumptions that may turn out to be false, for example, that the embryo/fetus would be easily transferable to an artificial womb. Nevertheless, the scenario does show that any new technology can have unexpected implications. In this case, abortion would probably become harder to justify. This alternative, pro-life choice would be an option to either death of the fetus or involuntary gestation by the pregnant woman.

BONDING

Critics will claim that ectogenesis harms mother and child because of lack of bonding. "What makes pregnancy special is the bonding that occurs between mother and child," says Bible studies professor Scott Rae.[10] But what exactly is bonding? It is the alleged biological connection developed between female gestator and fetus during nine months of pregnancy, but its existence is largely psychological speculation.

Many people generalized from the one case of Baby M to all of commercial surrogacy, and eighteen states outlawed this practice. But in California, commercial surrogacy has been going along quietly for nearly two decades. Anthropologist Helena Ragone enlightened us on this topic by actually interviewing women who had worked as surrogates, many of them from California.[11] Surprisingly, most already were mothers and saw themselves in altruistic ways. The money allowed them to work in (what they saw as) their natural, God-given role: to create life, to be mothers, to help others.

The other surprising result was that the surrogates didn't mind giving up the baby. The surrogates were sad at the ending of their role of pregnant mother and the feeling of being special to everyone because of it. This sadness confirms a widely reported result discovered by nurse-researcher Nancy Reame, who interviewed (only) ten surrogates who had given birth a decade before. Six of the ten expressed some disappointment, not at having been surrogates but because the "relationship had been abandoned by the adoptive couple at the time of birth (for 3 women) or over time (for 3 women)."[12] The disappointed six expected long-term contact with the adoptive couple—a continuation of their feeling of being special to the couple as the surrogate—but this was unrealistic.

Historical evidence also argues against bonding. Aristocratic women used wet nurses to breast-feed their newborns. If such aristocrats had bonded with their children, wouldn't they want to breast-feed them? If bonding were real, how could they so easily give away their babies for breast-feeding to a stranger?

One final point. First, notice how much work is done in reproductive ethics by appeals to psychological harm. Whether we are discussing cloning or artificial wombs, critics frequently appeal to this quasi-empirical, secular claim to justify their opposition to a new tool to help babies. Second, we should not accept the equation bandied around of late that feminism equates with opposition to biotechnology. Just because Judy Norsigian can justify abortion but not embryo research does not mean she speaks for most women, especially most infertile women. When it comes to overcoming infertility, most women embrace the latest biotech tools.

NONTHERAPEUTIC RESEARCH?

If we are ever going to develop an artificial womb, research will have to be done. But how will that happen? Obviously, if we can't use federal funds to experiment on human embryos, we won't be able to do so on human fetuses in such an environment. Thus the research will have to be privately funded.

For thirty years, the viability of a premature baby's lungs has been the absolute barrier to progress toward an artificial womb. Reports of successfully using liquids to substitute for the nutrients supplied by the amniotic fluid have been exaggerated. Until the infant can breathe outside the mother's womb, even with tiny respirators, it is not really viable. Any real research in this area would have to attack this problem.

Could such research go forward in private companies right now? Researchers in assisted reproduction clinics can now use human embryos to do research. No law regulates how private clinics can study human embryos, although no federal funds can be used for this purpose. Consequently, could private researchers try to gestate human embryos in extracorporeal gestation *right now*?

THERAPEUTIC RESEARCH WITH EXTRACORPOREAL GESTATION

The logical candidates for therapeutic research would be premature babies lacking viable lungs, probably aged 20–22 weeks or less. Such research would probably use a liquid environment mimicking the nutrients, oxygenated blood, hormones, and antibodies of the mother's womb. Like ventricular-assist devices that bridge to heart transplants, such extracorporeal wombs would bridge to NICUs and their tiny respirators. Hence the most likely place to start research on extracorporeal gestation will be as a therapeutic measure for dying premature babies. ECMO had to be tested and proven in just this way.

In the present political climate, it is inconceivable that government would fund such research. No administration in the past decades has funded research using human embryos, so you can imagine how controversial research on the artificial womb would be, which would involve third-trimester human fetuses.

It will be very difficult to do the practical research necessary to prove extracorporeal gestation safe for human babies. Critics will revive the old claim of fundamentalist theologian Paul Ramsey that such research would be "unconsented-to" research on the unborn, repeating the oft heard, nonsensical demand for consent from the subject for research before it is born.[13]

They will correctly say that ectogenesis research subjects a fetus/baby to an experiment lacking his consent and lacking well-characterized risk. But what Ramsey failed to mention is that every advance inside our children's hospitals subjected babies to unconsented-to risk, often without well-characterized risk. It is the nature of brilliant innovation that not all risk can be understood in advance.

ANIMAL MODELS AS INTERMEDIARIES

Long before an operative, efficient mechanical womb is available, animal models may be tried in last-ditch efforts. In some ways, a primate is a primate is a primate, and primate surrogates might be possible for some human fetuses, especially if the animal hosts were genetically modified to have immune systems compatible with surrogacy.

Jumping ahead to the next chapter about transhumans and transgenic animals, I note that a chimpanzee mother might be created with a few human genes to prevent such an at-risk human fetus from being rejected, in the same way that researchers are trying to modify pig organs so humans won't reject them. In such a case, the best interests of the child would run up hard against a current taboo in our society: no human fetus should be gestated inside a nonhuman primate mother. But if this were in the best interests of the child, wouldn't it be right thing to do?

SOCIAL JUSTICE AND THE ARTIFICIAL WOMB

Predictably, a safe, efficient artificial womb would be critiqued as creating a further divide between rich and poor, as rich women would no longer need to bear children. Just as some rich models (e.g., Cheryl Tiegs) hire surrogates to bear embryos created from their eggs to avoid changes in their bodies, so rich women in the future would simply deposit their embryos in incubators and go back to their normal lives. But what about poor women or women who lack medical coverage for this?

The generic answer is that we do not need to have a definitive answer to questions about distributive social justice to know that a thing is good or bad. Every innovation in electronics, medicine, pharmacology, aviation, or mechanics is potentially available more to people who can afford innovations than to people who cannot. The point of scientific progress is not to be a handmaiden to those who wish to create a perfectly egalitarian society.

More practically, because of the great benefit to the child, it is likely that medical insurance would indeed pay for artificial wombs. Although future savings from preventive medical interventions are often exaggerated, in the case of the artificial womb, this might be true. Especially if we consider benefits to fetuses/babies not exposed to alcohol, nicotine, or cocaine, and especially if we consider that such exposure is more common among low-income mothers than high-income mothers, then the artificial womb might not be a tool that created more social justice but the reverse.

Finally, the egalitarian objection is an objection about distributive justice, not about what is good or bad. Indeed, and as argued before, to even raise the objection implicitly assumes the thing distributed is good, such that it is wrong for some to have it and not others.

PREDICTABLE ALARMIST CRITIQUES

If an artificial womb was successfully developed, or even if scientists began to study it, it is easy to imagine the hysterical jeremiads it would elicit: an assault on the dignity of humanity (especially the "dignity of motherhood"), a dehumanization of childbearing, an affront against God's will, an attack on the traditional family, a psychological trauma for any baby born this way.

And, of course, few critics could talk about the artificial womb without mentioning the slippery slope. If gestation is made predictable and if humans are no longer needed for it, what is next? Ordering babies out of a J. Crew catalogue to be raised in an Orvis incubator delivered nine months later to rich parents on the other side of the planet? (It's fun to turn alarmist and see how easy it is to raise such specters!)

One can imagine the artificial womb being held up as a reductio ad absurdum of technology out of control rather than as a benefit to women. Critics, predictably, would actually say that women would be harmed by the artificial womb by being deprived of the benefits of gestation, such as bonding and (ahem) "the deep metaphysical joy of fulfilling their biological function."

CONCLUSIONS

Even the possibility of ectogenesis will alarm people, and conservative critics will predict the worst outcomes from this practice, which they will never accept as safe. Nevertheless, under carefully chosen circumstances, ectogenesis might be in the best interests of a fetus and/or the woman who otherwise might have tried to gestate the fetus. Research may be most justified on dying premature fetuses for whom no other options are available. Other objections to ectogenesis, involving bonding or just social allocation, are premature and speculative.

Even though to many people research on artificial wombs seems preposterous, a little reflection shows that pro-life constituencies might even welcome this technology. Given that reflection can turn around our intuitions about such a technology, perhaps there is hope that the same can be done one day for reproductive cloning. Let us hope.

15

Humanzees, Transhumans, and Transgenic Animals

> [Creation of GloFish] is the tipping point, when the world irrevocably turns toward the science-fiction fantasies of writers like Phillip K. Dick and William Gibson, who envision biomedical technology permeating every corner of the marketplace, from global corporations on down to small-time illegal operations like stolen-car chop shops.
>
> —*James Gorman,* New York Times, *December 2, 2003*

Artificial wombs and beings composed of human and nonhuman animal genes confront us with new ways of making sentient beings. Discussion of these ethical issues is not only interesting in itself but also sheds new light on cloning. Technologies to clone embryos allow insertion of genes to improve on nature. New kinds of transgenic animals can also be created for food, clothing, and research. Are these biocreations perversions? Good?

Creation of these new beings seems to be intrinsically wrong based on our initial emotional responses, but reflection yields some cases where each of these might be valuable and hence, morally permissible. Although each seems to be a border crossing fraught with danger, I shall argue that we should base our ethical judgment not on the act of crossing the border or its novelty, but on the characteristics of the new beings created.

Exploring the ethics of animal cloning and transgenic animals involves not only morality but also axiology, the general study of values, particularly axiology toward animals. In ethical theories that grant moral status to animals, this axiology raises distinctly moral questions.

OLIVER THE HUMANZEE

For over four decades, a remarkable creature named Oliver has existed among humans and caused considerable dismay among people who have met him. He has been the subject of documentaries on cable television and of national news reports.[1]

For a chimpanzee, Oliver possesses several remarkable characteristics. First, he does not scoot along on all fours but, like a human, he walks upright with strong arms held back and knees locked. Second, he has a remarkably humanoid face with an expressive mouth, a rounder cranium, prominent forehead, unusual ears placed higher on the skull than on most chimps, and inquisitive eyes. Compared to normal chimps, he has less hair, a flatter nose, and a smaller chin. He does not smell like a chimpanzee but like a human, so much so that other chimps avoid him. Remarkably intelligent, he learns things easily. Socially, he prefers the company of humans to chimps, sits in chairs, and once lived in a suburban house with a couple who owned a circus.

Onlookers speculate that Oliver is a humanzee—a cross-breed or hybrid produced from mixing chimp and human gametes. During his early years, word spread that geneticists hired by the circus had determined that he had forty-seven chromosomes, one more than humans, one less than chimpanzees. People also speculated that Oliver might be another kind of hybrid, perhaps between a chimp and a bonobo. Or a chromosomal break or genetic mutation could have caused Oliver's unique traits.

The happiest time of Oliver's life was when he lived with the owner of a circus and his wife. But Oliver became sexually aggressive toward the wife and was sold. Then he became an exhibit as a freak in circus sideshows. Worse, he was next sold to the Buckshire Institute in Pennsylvania, a laboratory that uses primates in research, where he was kept in a small five-by-seven cage for nine years.

In 1996, at age thirty-five, Oliver was tested by geneticist John Ely, who pronounced that he was not a hybrid but a 100 percent genetically pure chimpanzee. Oliver was a rare chimpanzee from central Africa as opposed to the typical chimpanzee from western Africa. Ely also said that some strands of Oliver's DNA were unusual.

Oliver and eleven other primates were then rescued and taken to Primarily Primates, a sanctuary for research chimps outside San Antonio, Texas, run by Wally Sweat. Sweat has seen Oliver walk upright and swears that chimps cannot be taught to do so. In June 2003, Oliver was forty-two years old, blind and crippled by arthritis, and no longer able to stand upright.

So Oliver wasn't a humanzee. Nevertheless, his life raises the question of how he should have been treated because many people around him *believed* he was. If a hunter stalking deer knows that children often play in a thicket, he shouldn't fire into it when he sees movement. People should act on their beliefs. Even though people believed Oliver to be part human, they treated him badly.

Some people think that a transhuman being like Oliver will be created one day, either through science, cross-species interactions, or mutations. If so, what should be the moral status of such a being? Should he be classified as subhuman and used for medical experimentation, or as a person with rights? Does giving a chimpanzee some human genes change his moral status? Does giving a human some chimpanzee genes demote the human in moral status?

We can easily spot parallels here with ethical questions about the moral status of cloned humans. In both cases, prejudice sees the new beings as beastly and subhuman, the property of human creators. Parallels with the artificial womb also emerge because people regard originating such new beings as paradigms of unethical acts by scientists (humanzee fetuses grown in artificial wombs? Sounds like *Island of Dr. Moreau*).

RECENT DEVELOPMENTS WITH TRANSGENIC ANIMALS

The biggest science story in decades, Dolly the sheep, born in 1997 by cloning, certainly upturned old notions of reproduction. But should every story of a new way of creating animals be reported this way? Since then, announcements about cloned animals have continued, along with news of genetically modified fruits and vegetables.

Such recent developments in moving genes around have produced the feeling in many people that scientists are creating biological earthquakes. Where previously mountains existed between species, now valleys exist with deep fissures. Future biological earthquakes might soon topple our ecosystem and, if nothing else, topple our sense of fixed moral boundaries between humans, animals, and plants.

For example, at Advanced Cell Technology (ACT) in Massachusetts, embryologist Jose Cibelli in late 1998 grew human nuclei using cow eggs that had previously had their nuclei removed. Although scant bovine ooycte material contributed only a few genes, and although the resulting embryo grew only to the very early stage when embryos with any defects fail to develop or implant (hence the resulting embryos were probably neither viable nor implantable in a human uterus), the yuck factor was so tremendous that a U.S. president condemned Cibelli's actions. Like the cloning of Dolly, the creation of a cow–human hybrid, even at the inchoate stage of an eight-cell unstable hybrid, seemed something out of *Dark Angel.*

In August 2003, Chinese scientists used enucleated rabbit eggs for human embryos in the same way that Cibelli did in 1998 with cow eggs.[2] Since Cibelli's work five years before had already produced the initial shock, which had been absorbed and mediated over the ensuing years, this time the reaction was muted. As with Cibelli's previous work, the rabbit egg used by Chinese scientists only contributed a tiny number of genes in mitochondrial DNA. The real import of the work was the attempt to create a biological minifactory to produce embryonic stem cells for research. Nevertheless, Richard Doerflinger, the Catholic activist for the U.S. Conference of Catholic Bishops, declared that the human–rabbit embryos were human enough to "deserve protection" (i.e., they should not have been created by such research). Alarmists such as Doerflinger make ordinary people wonder whether we have a fixed, static world ordained forever by God or a fuzzy, changing world created by humans for their own needs.

Citizens had also heard of a mouse bred to grow a human-looking ear on its side (to benefit humans who had lost ears), a geep (a mixture of goat and sheep), and Harvard's patented OncoMouse, genetically engineered to develop cancer tumors under certain conditions and therefore an ideal for cancer research. (Europe has balked at extending patent protection to the OncoMouse, as it opposes all patents on life forms.) Science also created mice with knock-out genes, allowing additions or deletions to study gene expression.

Chicago artist Eduardo Kac became famous when he hired French scientists to create Alba, a genetically modified rabbit that glows green under blue or ultraviolet light. Kac had French scientists insert a green fluorescent protein (GFP) into the original rabbit embryo. Although he enjoyed the fame, Kac was surprised that the French wouldn't let him take his rabbit home to America (because French authorities were aghast that the rabbit had been created there without public discussion). Kac is regarded by some fellow artists as a "mad scientist-artist." Kac claims his goal was merely "to know what it's like to live with a transgenic being."[3]

TRANSGENIC PETS

In late November 2003, a company called Yorktown Technologies announced it had added a gene to traditional zebra fish that made them glow bright red under normal light and fluorescent under ultraviolet light.[4] Dubbing them "GloFish," Yorktown Technologies hoped to sell GloFish in 2004 in pet stores.

Predictably, environmental groups opposed the sale of GloFish. The Center for Food Safety (despite its innocuous name, one of the most strident of contemporary environmental groups dedicated to egalitarianism and radical environmentalism) tried to delay entry of GloFish into pet stores, fearing that people would dump their aquariums into natural waters where GloFish might thrive.

This criticism carried some irony because GloFish were created in Singapore with the goal of being the marine equivalent of a canary in a coal mine. When the target fish encounter dangerous levels of pollutants in water, the idea is that they should glow. To develop that kind of fish, researchers had to prove that something could be done to make a zebra fish glow. Hence the GloFish.

Genes for biofluoresence were first added to tobacco plants and carrots in the 1980s and were safely added to mice in the 1990s.[5] So there is a record of safety. Moreover, the fish are genetically hobbled to prevent their living or propagating in the wild, something that has already been proven with zebra fish sold in pet stores, which cannot survive in the wild.

Of course, the world of botany and animal husbandry has seen crossbreeds for millennia, from the mundane world of roses to the more exotic plucots, a hybrid of a plum and apricot, which now graces our grocer's shelves. As we saw in the chapter on animal cloning, mules were created from horses and donkeys as superior pack animals. Researchers are also working on a cat that can live with humans and create no allergic reactions (chapter 4). That would be a boon to millions of cat lovers.

THE QUESTION OF EMOTIONAL SHOCK

Such different beings shock our sensibilities, a fact known to patients with neurofibromatosis (Elephant Man's disease), Little People (dwarfs), intersex and transgendered humans, and any other outlier, marginalized person who deviates from human norms.

In ethics, what value should be placed on our emotional revulsion to such new things? To the yuck factor? Probably very little. History shows that people disliked many things in the past on irrational grounds: Irish, Asian, and Italian immigrants, interracial marriage, racial integration, people from the Middle East, and any unusual practice, such as vegetarianism or nonsmoking.

Of course people worry about what Bernard Rollin calls "rampaging monsters."[6] Science fiction has primed us to worry this way. Moreover, our most ancient fables (see *Bullfinch's Mythology*) warn us of the dangers of mixed-species creatures such as sphinxes, centaurs, and griffins. Nevertheless, all this explains our emotional reaction to new creatures; citing such feelings is not a reason against creating such creatures.

ARE SPECIES ESSENCES?

Traditional hybrid animals were made by random mixing of genes in sexual combination. Some methods of gene insertion that create genetic hybrids randomly insert a few new genes, which means that scientists can control where they go and what they do. They just have to wait to see what the grown animal is like; if it's valuable, they let it reproduce sexually.

Knockout techniques allow a much more precise insertion of genes with known function. The same techniques are used in genetically modified vegetables and crops. Rather than messy, random, unpredictable combinations of hundreds or thousands of genes, new techniques allow for safe, accurate insertion of genes across species. This raises the question of whether a cow with a few goat genes is really much different than a traditional cow, especially because traditional cows contain variance in their genetic structure.

To make an analogy, a man with a pacemaker is certainly not a cyborg, nor is a woman with a plastic cornea or a metal hip joint. Nor is someone a cyborg who needs a machine to live, such as a dialysis machine, ventilator, or feeding tube. In these cases, we would say that the part or machine enables our humanity. Just because part of me went wrong (a heart valve) and was replaced with a machine or animal part (metallic or porcine valve) doesn't affect my human status at all. Similarly, having a few new genes in a being with thousands doesn't seem like a big deal.

What is the source of the fear of cyborgs? One answer is to look at the enduring power in our culture of Mary Shelley's distrust of science and her seminal novel, *Frankenstein.* What child or adult has not seen this movie, or versions of it? Part of the reluctance to move genes around between humans, animals, and plants is the belief by many people that each species is fixed by a common genetic essence. To contaminate such essences is to mess with nature, which left alone can work well.

Part of the opposition to transgenic animals and reproductive cloning rests on simplistic ideas about genetic essentialism. People think we are made of a dozen genes and to mix one or two or three would destroy the human essence and create something like griffins, chimeras, or centaurs. Charles Darwin, the man who discovered evolution and championed its working, rejected this view:

> I look at the term species, as one arbitrarily given for the sake of convenience to a set of individuals closely resembling each other, and that it does not essentially differ from the term variety, which is given to less distinct and more fluctuating forms. The term variety, again, in comparison with mere individual differences, is also applied arbitrarily, and for mere convenience sake.[7]

A review of the biological literature by Roberts and Baylis finds twenty-two definitions of species.[8] Similarly, Harvard professor Richard Lewontin asserts that no genetic essence exists that is shared by all humans.[9] "In sum, even though biologists are able to identify a particular string of nucleotides as human (as distinct from, say, yeast or even chimpanzee), the unique identity of the human species cannot be established through genetic or genomic means."

Similarly, modern evolutionary biology rejects essentialism, which dates back at least to Aristotle, because what we call a species is merely a rough, stable cluster of properties, not all of which are always necessary for a being to be counted as a member of that species. In the literature of philosophy of science, this is called species-concept antirealism.[10] Humans share no fixed essence, although people once thought this way. No subessences divide us into different races. Just as a continuous, fluid world molecular genetics threatens some worldviews, it enlightens others.

XENOTRANSPLANTS AND CROSS-SPECIES TRANSFER OF VIRUSES

Scientists hope to cross species barriers in practical ways to help humans. The Roslin Institute in Scotland, which has already cloned pigs, hopes to create pigs with human genes to create organs that can be transplanted to humans.

It is informative to compare the ethics of such intended xenotransplants with the ethics of human reproductive cloning. The great danger of xenotransplants is that they might introduce new viruses against which humans have no defenses. In many ways, and despite the hype in the press about its potential to help humans needing organ transplants, xenotransplantation is a hundred times more dangerous than reproductive cloning. Despite this fact, research on xenotransplants continues in many labs around the world, while research on reproductive cloning has been halted. Even embryonic cloning is at risk in America.

Why is xenotransplantation so dangerous compared to cloning? One, maybe two or three human babies might be originated by cloning, and if they are abnormal, the world will react and ban such cloning for a century. On the other hand, a new virus introduced by a xenotransplant gains a portal into the patient receiving the transplanted animal organ *and into the entire human race.* This is what occurred with the HIV virus and probably the SARS virus. Such a danger means that one patient cannot consent for us all to run such a risk and it is not clear how one knows, if ever, that such a transplant could be either safe to try or safe to commonly practice.

This gives us a few tips in thinking about ethics: first, the really big issue is the one you don't see, not the one that blazes each night on television. Because four thousand Americans die every year waiting for an organ transplant, demand pressures physicians to try more xenotransplants, for example, to repeat the baboon heart transplant done on Baby Fae in 1984. If our ethical dam breaks because of such pressure, a much greater flood may come on us than cloning.

We must be careful that the food industry is not creating portals for diseases to cross natural barriers and infect new species. When livestock in England were fed the dead remains of sheep, spongiform encephalopathies jumped to cattle and then to humans as mad cow disease, via newly discovered infectious proteins called prions. The real caution with transgenic animals must be in handling them as pets, eating them, and using them as medical resources.

Here we again have a problem of scale: a few babies created through reproductive cloning versus a risk of a lethal virus to the entire human race. As SARS and HIV have shown us, we don't really understand how such viruses get created or jump from animals to humans.

To understand an issue in bioethics, follow the money trail, such as why Dolly's birth was announced seven months after it happened (to allow the Roslin Institute to have its patents approved and its future income assured). Some critics who have made criticizing human cloning almost a full-time job have never mentioned the danger to humanity of lethal viruses from pig organ transplants. Why harp on the sensational danger of a few cloned kids while remaining mute on the dangers to humanity posed by basic research going on in most medical schools?

THE QUESTION OF MORAL STANDING

Our initial reaction to hearing of or seeing such a creature tends to be an emotional or intuitive sense that creating such a being is just wrong. A strong intuition, maybe universal, is that—whether God or evolution made them that way—species are just fixed and should stay that way. The typical reaction in public policy is to ban such creatures or inflict moratoriums based on emotional responses.

R. Albert Mohler Jr., president and professor of Christian theology at the Southern Baptist Seminary in Louisville, Kentucky, puts the Christian and theistic opposition to transgenic animals succinctly:

> First, the acknowledgement of our limited dominion [from God over animals] should make clear that our rulership is limited. We are not to take the authority of the Creator as our own. Stewardship and dominion of other creatures is to be exercised within limits imposed by the Creator. . . . Put bluntly, we were not commanded or authorized to create new forms of life as extensions or our own design and egos. . . . Human beings are assigned responsibility for the care, use, and enjoyment of animal creatures, but we are not granted license for their mechanistic manipulation, transgenic innovation, or ruthless violation.[11]

Answers to this question take us far beyond the merely practical question of whether certain kinds of bioscience research should not be done. Answers to this question are deeply intertwined with answers to interesting questions about which beings have, and which beings lack, moral standing in our evolving moral universe.

NO BORDER CROSSINGS?

In a groundbreaking article in the *American Journal of Bioethics* on the ethics of creating transgenic animals, Jason Roberts and Françoise Baylis argue that we should "avoid any practice that would lead us to doubt that humanness is a necessary (if not sufficient) condition for full moral standing."[12] They conclude that mixed human chimeras and embryos should not be created because to do so would add to the confusion about who is a human. If their "no crossings" argument were valid, it would also be an argument against creating cloned embryos or cloning to produce babies. But it is not.

Unfortunately, as many commentators in the same issue also conclude, they do not present an argument for this conclusion but rather an assertion. That assertion is that "the separateness of humanity is precarious and easily lost; hence the need for tightly guarded boundaries." Both the premise and the conclusion from it are tenuous and controversial and, as such, this argument is invalid.

Their conclusion ignores a quarter century of work in ethics that challenged as speciesism the privileged moral standing of human animals. Peter Singer, Bernard Rollin, and many others have argued that some humans are not persons with full moral standing (anencephalic babies), whereas some nonhuman animals should be given full moral standing (great apes, dolphins, and whales).[13]

MORAL STANDING AND SPECIES MEMBERSHIP

Put simply, the question that Roberts and Baylis beg is this: why does membership in a species morally count at all? Roberts and Baylis wrongly assume that the shaky boundary between human and nonhuman animals is worth preserving. But if the boundary is like skin color, gender, sexual attraction, or age, then it should be neither necessary nor sufficient for moral standing but jettisoned: keep good boundaries, lose bad ones.

One way of looking at the key issues in bioethics in the past three decades is that it has been addressing the master question of who is a person and who is not, which boundaries are expanded and which contracted. If we draw a series of concentric circles expanding outward from a central point, then great consensus exists that adult humans with normal cognitive faculties (the great bulk of humans on the planet) are humans with full moral standing. In this picture, some people want to contract the margins of conventional humanity by excluding as full moral persons fetuses, patients in persistent vegetative states and late Alzheimer's disease, and anencephalic babies.

Other champions want to expand circles of full moral personhood by including dolphins, primates, and a whole line of sentient creatures. Indeed, radicals in environmental ethics such as Charles Taylor would expand full moral standing to almost any creature who has a life, and ever more radical theorists claim that ecosystems have full moral value. Thus the question of moral standing connects directly to the question of our obligations. If a dolphin is just as much a person as my sister, then it is just as wrong to kill a dolphin as my sister, and the dolphin should be as legally protected against murder as my sister. As Hilary Bok writes, "In either case chimeras do not introduce confusion into our moral views. They reveal ways in which those views are inadequate and make us think about how we might improve them."[14]

Thus creating intraspecies hybrid animals might attack speciesism. People could no longer assert dominance of humans and inferiority of nonhuman animals; lines would blur as chimpanzees argued with their captors. Far from demeaning animals and making them into interchangeable resources, transgenics could liberate them. Interestingly, and according to Roger Williams law professor John Kunich, at present there are no federal laws barring creation of a mixed human–animal chimera.[15] Although such a creature cannot be patented, creating it is not a federal crime.

TWO EXTREME POSITIONS

Should we permit creation of humanzees and new mixed-species gene animal hybrids? One extreme position on this issue could be called laissez-faire. Perhaps the mercantile connotation associated with this phrase is appropriate because on this view, transgenic animals and plants are nothing more than materials and tools that can aid humans, created from capital invested in biological sciences in expectation of profit and should be the subjects of patents.

On this view, when such plants and animals create proteins and antibiotics for use in humans, they should be tested in clinical trials supervised by the Food and Drug Administration. But before that point, the FDA has no legal authority over the raw materials of biology, nor does the U.S. Department of Agriculture. No one should tell companies the kind of animal or plant they can create.

The opposite view might be called the Garden of Eden view, with its religious connotation that God created fixed, immutable essences of species and scientists are evil who open the Pandora's box of transgenics. The "no crossings" view of Roberts and Baylis, while not explicitly religious, falls in this category. On this view, vegetables are not meant to have genes of chicken in them, nor are chickens and goats meant to contain human genes. On this view, all transgenic animals are perversions of nature. Nature is not raw material to be species-bent to human needs, but something we should respect, not change, and live in harmony with.

Both views suffer problems. The laissez-faire view fails to explain how humanity is going to be protected from lethal viruses transmitted from animals to humans. In mad cow disease, we don't know how a protein is infectious, and deer and elk in northwestern America currently suffer from a similar encephalopathy. Until we know more, it's unwise to eat meat from such game and it seems especially unwise to permit more xenotransplants until we have a way to prevent transmission of viruses and such proteins.

The Garden of Eden view also suffers problems, perhaps more. As already noted, perhaps 20 million Americans are allergic to cats. Scientists have recently discovered that 60–90 percent of them are allergic to a single protein—Fel d1. In 2001 researchers at Transgenic Pets, a biotechnology company based in Syracuse, New York, isolated and sequenced the gene for Fel d1 and are creating a way to knock out this gene in a new allergen-free cat.[16]

It's hard to see how creating such cats would be bad thing. Many more people could have cats in their homes and life would be much easier for those with allergies who must now endure coexistence with such cats because of being married to, or a child or parent of, a cat lover. Ditto fish that glow in the dark while some industry fouls the river in the dead of night.

Much more important than transgenic pets, transgenic medicine promises to help millions, maybe billions, of people worldwide. In biology, *transgenic medicine* has moved along rapidly, with mice, chickens, goats, and even plants serving as biofactories to grow antibodies to fight disease. Many antibodies that fight disease aren't produced by sick people in the quantities they need to regain health. Sometimes such antibodies are difficult to grow in fermentation vats.

For example, Embrel, a miraculous drug for patients with severe rheumatoid arthritis, is so difficult to make and so scarce that it is given out in a lottery.[17] Yet it could be grown in plants in vast quantities if scientists were allowed to do so. Such *pharming* is more efficient in many cases than creating proteins and antibodies in traditional labs. Similar efforts are underway in greenhouses in California by Epicyte, a pharming company growing antibodies needed by humans in tobacco and corn plants.[18] Another company in Massachusetts has created similar proteins by adding a few human genes to goats, which secret the protein in their milk. Similar efforts are underway to create proteins in transgenic chickens and mice. Other companies are testing goats that grow vaccines, corn that grows insulin, and genetically enhanced crops for the Third World that contain vaccines against common deadly diseases of poor people.

Rather than being scarce and expensive, such transgenic animals and plants offer the real likelihood of production of desirable proteins in metric tons. All of these developments are a real boon for humanity and shouldn't be banned.

Watchdog and skeptic Sheldon Krimsky of Tufts University, a perennial critic of for-profit bioscience companies, agrees that putting one or two genes into goats or mice to make medical proteins "isn't a problem."[19] On the other hand, "There is a point, and I'm not sure where it is, where we shouldn't be putting in large quantities of human genes into animals." That certainly seems right too. So the Garden of Eden view wrongly blocks new drugs in transgenic medicine and new kinds of pets for humans, whereas the laissez-faire view falls to objections about safety to humanity from transgenic viruses.

Finally, there may come a day when a new kind of AIDS/SARS virus threatens humanity and we have no other defense than to inject DNA into human embryos from animals immune to this new virus. When it happens, we will want all the tools on the table to fight it, not some locked away in deep freezers that were never developed.

RACHELS AND DARWIN ON ANIMALS

In a book that deserves more attention, *Created from Animals: The Moral Implications of Darwinism*, the late James Rachels attacks the common notion that the only moral implication of Darwinism is to dethrone creationism and replace it with amoral evolutionism.[20] Instead, Rachels shows that Darwin saw a vast continuity—from humans through apes and chimps down to finches and worms—of intelligence and moral qualities. In his own time, Darwin drew conclusions that threatened humans and seemed to bring them down to the lower level of the beasts. But Rachels argues that this view, while somewhat true, is misguided because Darwin's real message concerns how other species of animal have also evolved to show social skills, cooperation, and intelligence. When it comes to qualities we prize, instead of equalizing humans down, Darwin was equalizing other species up.

Today we know more about animals and their behavior. We have overturned some past misconceptions. Peter Singer's *Animal Liberation* and People for the Ethical Treatment of Animals (PETA) have changed our attitudes about animals in the wild and their true qualities, as have countless shows on cable television.[21]

Our inherited attitudes toward animals and our newfound sensitivity about animal rights is the pedigree we inherit for thinking about transgenic animals and humanzees. But even our contemporary attitudes toward animals are not uniform. As Rachels beautifully explains,

> It has always been difficult for humans to think objectively about the nature of non-human animals. One problem, frequently remarked upon, is that we tend to anthropomorphize nature and see animals as too much like ourselves. An opposite but less frequently noticed difficulty is connected with the fact that, even as we try to think objectively about what animals are like, we are burdened with the need to justify our moral relations with them. We kill animals for food; we use them as experimental subjects in laboratories; we exploit them as sources of raw materials such as leather and wool; we keep them as work animals—the list goes on and on. These practices are to our advantage, and we intend to continue them. Thus, when we think about what the animals are like, we are motivated to conceive of them in ways that are compatible with treating them in these ways. If animals are conceived as intelligent, sensitive beings, these ways of treating them might seem monstrous. So humans have reason to resist thinking of them as intelligent or sensitive.[22]

For Rachels, we often misconceive the moral status of animals in two ways: first we tend to romanticize them and attribute to them human feelings, thoughts, and desires. Second, because we exploit them, we excuse our behavior by thinking of them as less than us and therefore as deserving the treatment we give them.

Generalizing from the two misconceptions to thinking about the moral status of transgenic creatures, we can predict that we will be prone to two kinds of mistakes: first, romanticizing animals and nature and seeing them as different than they are and, second (and perhaps more likely and certainly more dangerous), exploiting new transgenic creatures because they are different, other, and genetically impure.

One example of the first mistake, the common view of *folk essentialism,* goes back to Aristotle and holds that each species has a common, fixed essence. Folk essentialism is one way of romanticizing nature. Some transgenic creatures will be seen as more novel than they really are because nature will be seen as more static than it is—having fixed, unchanging essences of species, rather than, as Darwin saw it, being a fluid, changing continuity.

Nature already contains an astonishing variety of fishes, birds, insects, and small mammals, but most humans don't come into contact with even a small percentage of them. So when we see fish glow under ultraviolet light or learn of new mice helpful in the study of cancer, we may think these creatures radically differ from what already exists, but that is false.

I shall include this mistake in a discussion of the permissibility of creating farm animals that suffer less in being created for our food. The folk essentialist thinks there is something inherently wrong about creating a pig that can't feel or celery that tastes like chicken, but perhaps sustained reflection will change his mind.

A better example of the first mistake is the sad tale of the death of Keiko, the killer whale whose story was told in the 1993 movie, *Free Willy.* Captured near Iceland in 1979 when he was two or three years old, Keiko was sold as an exhibit in a Mexican amusement park. The Humane Society of the United States decided to enact a Hollywood-movie style rescue of its own, acquiring Keiko and trying to train him to kill food on his own and live among other killer whales. Released in the summer of 2002 into the Northern Atlantic near Iceland, Keiko showed up in September off the coast of Norway and entertained locals by performing his tricks. Loving to be around humans, he never learned to be a proper killer whale and shouldn't have been made to go against his real, adult nature. He died of pneumonia a few months later.[23]

The second mistake is to treat beings slightly different from us as resources for exploitation. Too many instances exist in history where humans infer that any difference in fellow humans justifies killing or enslaving them. One thinks of the *Star Trek* episode that featured a group of humans, half of whom have faces white to black from left to right; the other half, faces white to black from right to left. Each human group wars on the other, trying to enslave it, believing the other pattern to be inferior.

THE BASIC QUESTION

The most basic question in public policy about humanzees and chimeras is a philosophical one: what is the ethical status of any being created of human genes plus animal genes?

A step in the right direction was taken by the director of the U.S. Patent Office, who said that he personally believed that no being who was a human chimera could be patented. This is a step in the right direction because to allow such beings to be patented would be to regard them as things, not people.

Our first guiding rule stems from awareness of past attempts to marginalize humans who act and look different. Couple that with existing expectations of cloned humans based on science fiction, and there is possibility of abuse. The conditions for personhood are not exact but form a bundle of conditions, none of which are sufficient. This *bundle theory of personhood* includes the conditions of being able to feel, think, interact with people, initiate actions (agency), be reflectively aware, and be conscious of the external world.

Notice that these philosophical criteria do not mention having a human body or a human genome. These criteria leave the door open that a computer could be a person, that a chimpanzee could be a person, or that a humanzee could be a person. These criteria place such a high value on cognition and interaction that an alternative name for this bundle theory might be the cognitive-interactive model of personhood.

Notice also that these criteria presuppose actual capacities, not just potential ones. Because human embryos do not think or feel, they are not persons. Rather, it is most correct to say that they are potential persons.

CHICKEN CELERY?

Suppose scientists can make living stuff that has all the nutritional benefits and taste of chicken but is not sentient, like celery. Like the celery plant, it could be grown in vast fields under controlled conditions for water, fertilizer, temperature, and light. Would growing chicken this way violate some natural order of things? Would it be a bad thing? In a world where God made each thing with a fixed essence, it would be wrong to create chicken celery. Doing so would violate the essence of both chickens and celery, as well as giving humans a godlike role in creating new animate beings.

On the other hand, assume that the billions of humans on planet Earth will continue to love to eat chicken and that demand for it will grow, as it has in the past decades, despite more people in advanced countries becoming vegetarians. In that case, creating successful chicken celery would eliminate the need to produce billions of chickens, which would never exist and thus never suffer to become food for humans.

Of course, one can doubt the premise and argue that chicken celery would never taste like real chicken, that gourmets would spot the difference, and so on. But if it were good enough to pass unnoticed as cubes in chicken soup or McNuggets at McDonald's, that would be a great advance. And because pigs are as smart or smarter than cats and dogs, the greatest leap would be pig celery, which would replace the smell, waste, and horror of the huge pig farms with thousands of acres of pig celery.

I can't think of an image artists and critics would more like to lampoon than fields of chicken or pig celery, and, considered emotionally, creating such food seems like a perverse violation of nature. But if the suffering of billions of pigs and chickens now raised for our eating pleasure were prevented, isn't that more important than our inherited moral responses? Don't the same inherited responses make us feel it is fine to torture and kill animals raised for food? Isn't thinking about all that part of what human reason is for?

CONCLUSION

The big mistake in thinking about creating transgenic animals and humans in new ways is to think that any new way of creating humans is wrong, but when it comes to animals, anything goes because they are our property and they lack moral status. This puts too little weight on the suffering of animals while at the same time elevates primitive methods of creating humans to metaphysical grandeur.

What revulsion to cloning and transgenic animals share is an underlying folk belief that we are messing with essences that should be permanently fixed, by either God or evolution. That is a popular mistake, caused by many things coming together, but still it's a mistake. Instead, we should evaluate each being not by how it was originated or by how far it departs from standard methods of origination or stereotyped views of species essences, but by the particular qualities of each being itself, especially when it is a *de novo* being.

If a transgenic mouse exhibits communication, social interaction, higher skills, and rational goal seeking, we should treat it as a person, not a normal mouse. This is not human rights for mice but rights for mouse/humans.

We shouldn't worry about new beings upsetting biology or evolution: these fields already champion nominalism or antirealism about essences. Biology, especially at the DNA and molecular level, is already much more fluid than most people realize.

We should worry, though, if genetic changes in animals make them and us more vulnerable to new infectious diseases such as SARS, HIV, and mad cow disease. Globalization makes a more connected world, especially in agribusiness and medicine, and we need no new portals for such diseases. On the other hand, new techniques in our biotechnology tool kit may allow construction of new biological firewalls against such diseases, with transgenic food animals as our first line of defense.

The default position should be to make any transgenic changes in animals for the benefit of the animal. If such new beings are sentient, they should be assumed to be capable of suffering and such suffering should be minimized. If transgenic animals indicate by their actions, reflexes, and interactions that they can reason, communicate, and socialize in ways similar to humans, they should be assumed to be humans with full moral rights. Their moral status should derive from the particular batch of characteristics they have, not their degree of deviation from some imagined essence or by their unique method of origination. We should see Frankenstein's creation as a unique creature and not persecute him as the local villagers did. How such beings are originated and how many human genes they have shouldn't matter.

16

Three Challenges to Cloning Great People

No delusion is greater than the notion that method and industry can make up for lack of motherwit, either in science or practical life.

—*Thomas Henry Huxley,* The Progress of Science

This chapter discusses three problems associated with reproductive cloning. Some of the implications of these problems have been discussed by others; most have not. In order to raise the following problems, I make a controversial assumption; if it is false, the problems disappear. That assumption is that some version of genetic essentialism may be true—that people created from ancestors of the genotype would be substantially like the ancestor, especially if raised in similar families and roughly similar educational systems. Thus a child from Napoleon's genes would also be a great military strategist and leader of men.

A less controversial version of the same assumption is that genetic essentialism might be more or less true, or more determinative than egalitarians believe. Even with this milder assumption, the problems below seem to arise.

BIOLOGICAL INEQUALITY

As I sketched in chapter 1, in the great debate over nature versus nurture, egalitarians fear that genes determine too many variations of abilities among humans, whereas social conservatives do not. Head Start, run by the federal government to improve the life chances of poor children, symbolizes this debate. Started in 1965, it bypasses states, funneling money directly to poor communities for medical and dental care for children, nutritious lunches, and basic social and academic skills. It has affected as many as 20 million American children in thirty-eight years of operation. But has it been a success?

Egalitarian Alvin F. Poussaint, a black professor of psychiatry at Harvard Medical School who participated in the first years of the program in Jackson, Mississippi, argues that Head Start should be expanded.[1] Richard Hernnstein and Charles Murray, white authors of the controversial *The Bell Curve*, think that Head Start should be abolished, arguing that inherited abilities account for differences in intelligence and success.[2] Murray even argues that Head Start has made minority children worse off by fostering their dependence on government aid.[3] In *The Nuture Assumption*, Judith Rich Harris argues that Head Start has made no difference at all in the lives of children it affected because their personalities and intelligence were determined by their genes.[4]

Compromises are also around. Psychologist David Reiss for twelve years studied 720 pairs of adolescents and twins. He writes in *The Relationship Code*, "Many genetic factors, powerful as they may be in psychological development, exert their influence only through the good offices of the family."[5] Whether a genetic disposition to shyness in a child is allowed to flower depends on whether parents encourage the child to be shy or take steps to make the child more extroverted.

Given this background, egalitarians regard any new biological inequality as especially pernicious. It is one thing to tolerate inequities in family support, early childhood education, and quality of K–12 school systems, but deeper kinds of inequality must be nipped in the bud, especially a metaphysical inequality, hitherto unknown among humans, that is written not in the shifting sands of unequal environments but in the cold stone of genes. Environmental inequality is tolerable in liberal democracies, biological inequality is not.

Fast-forward to the time when reproductive cloning becomes safe. For egalitarians, safe human cloning would allow biological dynasties. To prevent that, egalitarians want reproductive cloning to be made illegal in all countries. Allied with the environmentalists described in chapter 15 who fear biotechnology, this new coalition is a powerful force on the worldwide stage against reproductive cloning for reasons other than safety.

Egalitarians fear a future society of billions of "Naturals" propagating primitively just as their ancestors did for millions of years, the bulk of their children regressing to the mean. Within that society, and increasingly leaving it behind, the "Enhanced" advance each generation, carefully controlling the reproduction of their children; only in the new genetic aristocracy, the "Enhanced" control their children for genes and ability.

Princeton genetics professor Lee Silver speculated about genetic classes in his *Re-Making Eden*, musing that if this process were carried out a thousand generations, a new kind of species might emerge, the "GenRich," so different from "Naturals" that the members of the "GenRich" could not interbreed with them.[6] Concluding his book, he speculates:

> In the twenty-fourth-century . . . humans would diverge into just two species—the GenRich and the Naturals. Naturals had the standard 46 chromosomes that long defined the human species, while the GenRich alive at that time had an extra pair specially designed to receive additional gene-packs at each new generation. With 48 chromosomes and thousands of additional genes, the GenRich were, indeed, on their way to diverging apart from the Naturals. . . .
>
> It was a long-sought-after genetic enhancement—finally perfected in the twenty-seventh century—that made it possible to even *think* about traveling to other solar systems. This was the gene pack—designed by Macingene—that slowed the aging process down to a crawl. Children born with the AGE-BUSTER gene-pack would live for hundreds of years, perhaps longer, with minds and bodies intact. Like young explorers throughout all the centuries of human existence before the twentieth, they said good-bye to their families knowing they would never see them again, and boarded enormous citylike nuclear-powered spaceships to travel to inviting planets discovered by astronomers in nearby solar systems.[7]

Do we really need to worry about this possibility? Do egalitarians have a legitimate fear that such biological stratification would create more social stratification? I think they do, for reasons described below.

WORLD ON FIRE

Professor Silver's musings about future "GenRich" and "Naturals" need not worry us because it is dozens and dozens of generations away, and hence not a realistic worry for us at present. On the other hand, what may really be a problem is the more legitimate worry described in *World on Fire* by Yale law professor Amy Chua, whose impressive argument has caught the attention of leaders of think tanks around the world.[8]

Chua attacks the happy-face picture of globalization of a rising sea of free trade creating markets all over the planet, in turn creating stable middle classes devoted to property rights and commerce. These middle classes in turn will create democracies, so sayeth the Happy Picture. With easy travel and intermarriage, in a few hundred years the universal man is Adam Smith with a happy brown face.

Professor Chua provides striking evidence that the picture is incorrect of a future composed of happy, middle-class, democracy-loving, market-friendly brown people across the planet. Her insight started brutally, with the murder of her Filipina-Chinese aunt in the Philippines by her chauffeur, an ethnic Filipino. In collusion with two ethnic Filipino maids, the murder was planned and executed. Surprisingly, the ethnic Filipino police did nothing to solve the crime, shrugging it off as justified by "revenge" for the injustice of Chinese wealth.

Revenge? Wealth? What's going on here? While the West has obsessed over its own concerns, great changes have been occurring throughout the planet. In the Philippines, "nearly two-thirds of the roughly 80 million ethnic Filipinos in the Philippines live on less than two dollars a day. Forty percent spend their entire lives in temporary shelters. Seventy percent of all rural Filipinos own no land. Almost a third have no access to sanitation."[9]

In contrast, and according to Chua, "just 1 percent of the population, Chinese Filipinos, control as much as 60 percent of the private economy, including the country's four major airlines and almost all the country's banks, hotels, shopping malls, and major conglomerates." They also own the best department store chains, major supermarkets, six of ten newspapers, the shipping and textiles industries, most real estate, and computer networks, as wells as businesses specializing in construction, manufacturing, and pharmaceuticals.

What we have here is a recipe for hate and resentment, as the indigenous peoples see a highly intelligent, hardworking, frugal, family-valuing, ethnically proud elite capturing the economic and human resources of their land. Chua's insight is that adding democracy to this situation does not produce stability because there is no stable middle class. Instead, politicians advocating social justice rally the masses to rise against their oppressors, using the ballot box to seize power.

As University of Virginia professor of political philosophy Loren Lomasky emphasizes, "Of all political structures human beings have devised for themselves . . . it is *liberal* democracy, not the pure sovereignty of majorities, that merits plaudits."[10] Liberal democracy with legal protection for minorities and individuals is needed in Iraq to prevent the majority Shi'ites from exploiting the minority Sunnis and Kurds, and it is precisely such protections that emerging democracies need to prevent (what Chua calls) a "world on fire" with populist democracies.

My concern here is not with stable democracy and globalism, but with how the pattern exists all over the planet. As Chua summarizes:

> Market dominant minorities can be found in every corner of the world. The Chinese are a market-dominant minority not just in the Philippines but throughout Southeast Asia. In 1998, Chinese Indonesians, only 3 percent of the population, controlled roughly 70 percent of Indonesia's private economy, including all of the country's largest conglomerates. More recently, in Burma, entrepreneurial Chinese have literally taken over the economies of Mandalay and Rangoon. Whites are a market-dominant minority in South Africa—and, in a more complicated sense, in Brazil, Ecuador, Guatemala, and much of Latin America. Lebanese are a market-dominant minority in Nigeria. Croats were a market-dominant minority in the former Yugoslavia. And Jews are almost certainly a market-dominant minority in post-Communist Russia.[11]

To Chua's list, we could also add the Indians from the Goa province of India, many of whom returned after the death of Idi Amin (who originally forced them to leave) and quickly jump-started and took over the economy.[12] The ethnic cleansing that Amin did to the Indians in 1972, which was initially popular with Ugandans, is seen by Chua as a possibility in Latin America, where indigenous Indians elect demagogues such as Hugo Chavez in Venezuela or Alejandro Toledo Manrique in Peru.

What does this have to do with cloning? The answer is: potentially a lot, although it will never make headlines. To understand this point, first, we need to ask how such ethnic dynasties have acquired, secured, and passed on their wealth, despite not being natives and not being respected by indigenous peoples. Since much of this is secret and private, much of the answer we do not know. We can infer, however, that such families exert great power over their children, both in forming their character and in shaping them to take over a family business.

Second, I have emphasized in this book the concept that both therapeutic and reproductive cloning are just tools and are not inherently evil; like all tools, they may be used for good or bad ends, for saintly or evil motives. Anecdotal evidence already indicates that wealthy ethnic minorities make aggressive use of assisted reproduction. Especially when their children pursue advanced degrees in finance, law, or medicine, they are likely to be at risk for infertility when they attempt to conceive in their mid- to late thirties. Consequently they have been using in vitro fertilization and egg donation to have children.

We don't have to worry about ethnic elites wanting to have children from the genotypes of Robert Redford or Meg Ryan. They will want their children to be of their own ethnicity, to blend in to their existing family structures, and to have genetic connections to their ancestors. Among the Chinese and Japanese, ancestor worship is a quasi-religion, and cloning ancestral genetic bases may become very popular as a respect to the founders of the dynasty.

Third, given the increasing competition across the globe, in part from other ethnic minorities striving for global domination, existing ethnic minorities will increasingly seek to shed the burden of children in their families with disabilities, retardation, or merely those who are ordinary. Chua's observations allow magnification of the egalitarian objection and a prediction: existing wealthy, talented ethnic dynasties will use safe human cloning to control the genotypes of their descendants. Rather than Hitler's Aryan master race, the real threat comes from those Hitler hated: Jews, brown-skinned Hindus, Chinese, and Lebanese.

By carefully controlling marriages of sons and daughters, these "Corleones" will promote their Michaels while controlling their Sonnys and letting their Fredos out to pasture. They will thus use selection of familial genotypes to amass and pass on wealth, social-political connections, and talent. Against such families, individuals from poor families will have little chance of competing for contracts, government grants, or funds for large-scale capitalization.

Anyone who thinks this is a real threat might reply that this is why the United Nations needs to ban cloning. But it is naive to think that a U.N. vote could have any effect on such powerful families. It is like condemning pornography while ignoring the irreversible changes in the way the Internet can make all kinds of pornography available to anyone. In a similar way, and once perfected, the techniques of creating and introducing a cloned embryo are likely to take place in private places, out of sight of nosy reporters or federal officials. There will be no satellites that spot couples cloning their own embryos.

Just as the rich and powerful use SAT coaches to improve their children's test scores, tutors to help them write lively essays, or specialists to help them volunteer for the right charities and projects, so too the rich and talented will one day use cloning to create children who are richer and more talented. One can bemoan this prediction, but I believe that little can be done to prevent it. John Rawls, the famous egalitarian from Harvard, tellingly writes in *Theory of Justice* that the family often contains the greatest sources of injustice, as it will funnel all its resources into advancing the success of its own children while spending virtually nothing on other children.[13]

As once said in a different context, "ought implies can." To which we might add, "ought not implies can," meaning that we can't say something ought not to occur, such as death, if we cannot prevent it. So two different questions arise: (1) *can* we prevent rich ethnic dynasties from using reproductive cloning? and (2) *should* we?

For someone trained in philosophy, at times it may be wise to take off the hat of the ethicist with his world of duties, obligations, and rights, and put on the more reflective hat of the philosopher. Rant as we may against more people driving cars, using cell phones, being manipulated by the mass media—these things seem inevitable. In a similar vein, given the drive of people to give their children advantages, given the threats to rich ethnic minorities of backlash and ethnic cleansing, and given their proven ambition and drive, is it realistic to think that safe reproductive cloning can be outlawed on the planet?

The process of enforcement would be formidable. Who would monitor cloning? How? Who would verify that a child had been cloned? How? Who would have the genotype of the ancestor for comparison? Could you compel her to donate blood or skin to test for a match? Will every country on the planet outlaw cloning? What would be the penalties for cloning? Who would be punished? The child? The ancestor? The parents? The family? The physician-scientists?

Cloned adults in these market-dominant ethnic minorities will be indistinguishable from any other adults and will likely fit perfectly in to large families. To protect the secret, the family may never tell them of their unique origins and may never have more than twins from one ancestral genotype.

I am referring to some of the most powerful families in the world, including WASP and Jewish families with old money in North America. These are families who know the personal numbers of senators and whose corporations employ professional lobbyists. When reproductive cloning becomes safe, when children of such families are infertile, do we really think we can prevent such families from lobbying to create legal ways to reproduce their genes asexually through cloning? Or if illegal in North America, finding another country to allow them to use cloning?

Nor is the center of biotech likely to remain in the United States. Because social conservatives have become powerful bio-Luddites in Europe and North America (see chapter 17, "Gathering Darkness"), we will lose the edge in both kinds of cloning to the Pacific Rim, especially China, which has already produced several cloned mammals and is using genetically modified crops to feed its billions. Singapore has spent over $2 billion to reinvent itself as a biotechnology hub, creating a biotech campus on 500 acres and setting aside nearly $300 million for scholarships for its students to pursue science doctorates at home and abroad.[14] Already it has lured pharmaceutical giants Glaxo, Eli Lilly, and Novartis to set up labs there. India is following suit, trying to re-create in biology the success of its great computer industry in Bangalore. Do we really believe we can prevent scientists there from creating, studying, and manipulating cloned embryos? Or studying them with an eye to fixing problems in reproductive cloning?

Why is voluntary, family-centered eugenics likely but state-coerced eugenics not? First, such families already have mastered control over their children's marriages and childbearing. Upper-caste Indian families in America still arrange marriages for their sons and daughters to upper-caste children with good prospects and good ancestors. Second, these families have already demonstrated the iron will to do what is necessary to achieve dominance in a world where they perceive that racists and fundamentalists hate their kind. The Chinese technological elite is not going to build America's railroads again and then be sent back in penury. Third, such families already heavily invest in the education of their children, often making sure they speak several languages and are tops in math, science, finance, and communication skills. Fourth, such families have the motives to use cloning in this way: they avoid the Fredos and maximize the genetic endowment of offspring.

Now comes the hard question: if I am correct about this prediction, does it make the egalitarian objection so potent that we should make cloning illegal both in North America and on the planet, such that such dynasties could only be created by back alley reproduction?

Even as a bioethicist, one can be philosophical. Market-dominant ethnic minorities secure wealth because they have expertise, capital, technical education, and they work hard; they make the mines, telephones, and big farms work. As Idi Amin discovered when he kicked the Asians and Indians out of Uganda, native Ugandans couldn't do it.

Earlier I emphasized that cloning will have no effect on the human genotype, given regression to the mean in population genetics. That is not to say, however, that cloning the genotypes of superior individuals in family dynasties will have no effect on the concentration of wealth and talent in market-dominant family dynasties in particular countries.

CLONING GENES OF GREAT PEOPLE

The first great problem for safe reproductive cloning is that it may be used as a tool to increase biological inequality and the concentration of wealth and talent across the planet in a few key families. A second problem has to do with cloning the genotypes of great individuals.

If we ever can safely re-create the genotype of a great person such as Lawrence of Arabia, we will be able to answer an interesting question that was the crux of Thomas Carlyle's theory of history. Carlyle argued that great men shaped history's big events, such as Genghis Khan, Napoleon, Alcibiades, or Adolf Hitler. But are great men great because their genetic nature compels them to rise to the top, or are they merely the right genetic base molded to the right phenotype and that appears at the right time in history in the right country? In short, would a great man be really great, again? Would the genotype of Napoleon be a great general or a financier, or just a megalomaniac who ends up in Leavenworth federal prison?

Actually, this question must be separated into two parts: first, would he be great, and, second, would he be good? These questions are not easy to answer, in part because of the relativity of good to a particular man's contribution to a particular society at a particular time and place in history. Take Napoleon. Was he good or bad for Europe?

The one true argument against cloning the genes of a modern Galileo or Leonardo da Vinci arises in the following way. Assume that cloning is just a tool of human origination, neither good nor evil in itself, and, second, that re-creating the genotype of an ancestor tells us nothing of how that new human would feel and act. Given these assumptions, the problem is that if we re-create genius, we cannot guarantee that the genius will use his talents for human good. Especially because, as I have emphasized in this book, the new person will have free will, we cannot guarantee any result from re-creating his or her genotype.

Because we know genetic essentialism is false, the new person will differ from the ancestor. A person originated from the genes of Gandhi could be violent, or a person from Mother Teresa, an indulgent narcissist. There are simply no guarantees. Fetal environment, childhood, free will—all are variables that might make a true genius turn from good to evil. A rebellious teenager could turn against his expected duty, forming his adult character in such resistance.

Producing a dozen progeny of a modern chemist as brilliant as Robert Boyle would be a gamble: we might get some dedicated to using chemistry to cleaning up the environment or making wonderful drugs to kill cancer, but we might also get others who created synthetic heroin that could be sold for pennies.

CLONING AND SPORTS

A third problem, much cited, concerns the effect that cloning the genotypes of great athletes might have on sports. Critics of biotechnology always recite the mantra that "technology is growing faster than our morals," and that "we should understand the ethical implications of technology before we implement it." An example that leaps to people's minds about human cloning is to re-create the genotypes of famous athletes such as Michael Jordan. Would the ability to safely clone humans change sport as we know it?

Perhaps. Any change forces us to think, and perhaps more deeply than before, about how we define a sport, the purpose of competition, and the rules of the sport. In this sense, cloning is a reductio ad absurdum of sport, and as with all reductios, it forces us to go back to the initial premises and rethink: What exactly is sport? Whom is it for? Sport involves, among other things, entertainment, achievement of human excellence, personal best, business connections, and simple human fellowship.

Looking at the Special Olympics may help here. Does use of prosthetics or mechanical aids mean that disabled people are not participating in sport? Of course not. They are merely participating in a different kind of sport.

Take Casey Martin and the Masters golf tournament. Should Casey be able to ride a cart? Is golf about hitting and putting a ball after walking and being exhausted or is it about hitting and putting a ball? If golf is primarily a social and business affair, then questions of admitting women and blacks become important.

What's understood as fair and unfair, natural and unnatural, equal or unequal, in sport has many layers of meaning, having to do with a sport's entertainment value, the flourishing of the individual, and the integration of people into society.

So perhaps the "GenRich" will only be allowed to participate in GenRich sports. If a college or professional athlete eventually gets a bionic eye, should he be banned? Even on a social definition of golf, one guy with a bionic eye will be seen as unfair. But is use of a protractor unfair in golf?

The GenRich Olympics would be a parallel Olympics for people originated in novel ways. In GenRich Olympics, there could be an algorithm that determines a competitor's score based on not only her time but also her level of enhancement. This reverses the way athletes are scored in the Special Olympics, where an algorithm determines the score based on the time in a race plus level of disability.[15]

Of course, the question of what is natural in sports is already under some stress from genetics. About one in five hundred people have some sexual anomaly that makes them suspect as a male or female. Humans with Y chromosomes and lots of testosterone look very masculine and may be accused of taking synthetic testosterone.

Environmental enhancements already stress the idea of natural competition in the Olympics: athletes from developed countries have million-dollar training gyms, swimming pool complexes, horse riding courses and arenas, special teams of expert trainers, coaches, physical therapists, and sports psychologists. No wonder we cheer for the underdogs when athletes from developing countries win. (Even then, does biology rule? The marathons are usually won not only by Kenyans, but by Kenyans from the same villages in that country, suggesting a tribe with rich genes for long-distance running. Should Peyton and Eli Manning be banned because they are lucky enough to get great genes from their father, Archie?

Cloning the genes of great athletes also raises the question of effort versus biology, a question that in some ways parallels the nature–nurture controversy that began this chapter. Some people say that it is one thing for the elite to send their children to the best schools, hire them academic tutors and personal athletic coaches, but at least the child must makes some effort to become talented. With a built-in biological talent, little such effort is required because of the tremendous natural advantage. It is like having all children run a mile race when the rich ones get to start a quarter mile ahead.

In one aspect, this debate resembles the one in neuroethics about the use of drugs versus more natural means. Therapy naturalists disdain drugs such as Prozac or benzodiazepines (tranquilizers), arguing that only discussion, feedback from a therapist, and insight can lead to real changes of character. They would hold the line with other drugs too, allowing a medication such as Aricept to be used to prevent the breakdown of the brain in Alzheimer's and Lewy body dementia but ban its use (current among some medical students and psychiatrists) as a memory-enhancing drug. But who is to say that equanimity gained by climbing a peak in the Rocky Mountains is any better than that gained by someone who goes from paralyzing anxiety before a test to a state of calm competence?

CONCLUSION

Market-dominant ethnic minorities creating biological dynasties to concentrate wealth and power in fledgling democracies sounds like a political powder keg for the United Nations and World Court. Ditto whether cloning genotypes of geniuses might give us not only more Gandhis but also more Rasputins. And cloning genes of great athletes might change amateur and professional sports as we know them or create some new, advanced kind of sports.

Do these problems, alone or together, constitute knock-out arguments against legalizing safe, reproductive cloning? I don't think so. As I argued earlier in this book, it's the problems you don't see coming that really bite you, not the ones you overanalyze. By shining a light on these problems, we can start a healthy discussion and see how serious they are.

17

Gathering Darkness or Transhumanist Light?

We are as gods and we might as well get used to it.

—*Stewart Brand,* Whole Earth Catalog, *1967*

We live in confusing times for students and the general public, with biotechnology under attack everywhere. Alabama Christians, feeling themselves under attack, elected fundamentalist judge Roy Moore as chief of the Alabama Supreme Court, despite his complete lack of administrative experience and qualifications, because he represented the sentiments that "Christianity is under attack" and that "religious values are not running public policy."

What's true in Alabama is also true nationally: according to a survey by *USA Today,* whereas 66 percent of Alabamians favored Judge Moore's efforts to put the Ten Commandments in the Alabama Supreme Court building, fully 77 percent of Americans nationally do (perhaps fewer Alabamians support Moore because we know him close up). The dominant culture increases power by telling its members that it is under attack by sinister, powerful forces ("foreign terrorists" and biotechnology).

Body bags from Iraq make George W. Bush vulnerable in the presidential election in November 2004, so Bush must court Catholics and Protestant bio-Luddites by protecting the embryo. Karl Rove has calculated that he can sacrifice the few infertile couples who might be harmed and gain votes among the vast number of elderly, who buy the "embryo/baby" slide. Leon Kass sees this too and is going for the big prize: restrictive federal control of fertility clinics.

As 2003 closed out, *Wired* magazine resurrected its old tricks with another sensationalized cover story, "The Making of a Human Clone."[1] The story, couched in the breathless hype of an insider at the Manhattan Project, told about . . . what? The creation of a sixteen-cell human embryo, not even a blastocyst, and certainly nothing sentient or capable of implanting in a woman's uterus. Moreover, Kyla Dunn had already done the ACT–human embryo cloning story a year before in the *Atlantic Monthly* magazine.[2]

I recently was privileged to hear some lectures by a professor-scholar about the life and films of Charlie Chaplin. Subsequently I watched *Modern Times* and *The Great Dictator*, as well as the superb film starring Robert Downey Jr., *Chaplin*. What is so sad is how Chaplin's J. Edgar Hoover helped suppress Chaplin's films in America, such that many of us have never seen them. What is even sadder is that Chaplin's *Great Dictator* was far ahead of his time, warning us of the dangers of this imitation "Little Tramp" at a time when America, in its isolationist coma, didn't want to hear about Hitler.

In *Chaplin*, in which Robert Downey Jr. plays Chaplin, Hoover is made to say something to Chaplin along the lines of "You'd better get smart with your movies. They influence how people in America think." Hoover disliked Chaplin's distrust of public officials and concerns for the workingman, branded Chaplin as a communist, and barred him from reentering America. Looking back, it seems incredible that this happened. Looking back, it is easy to see how it happened. Hoover held power unchecked for fifty years. Joe McCarthy swelled to excess and also tried to control Hollywood and its films with his famous blacklist.

In the past century, movies and television became the battleground for people's minds. In the coming century, biotechnology will be a new battleground over people's bodies.

GATHERING STORMS

In 2003 the Hastings Center, a bioethics think tank, published a report calling for federal regulation of (what it called) "reprogenetics."[3] A few months later, the Kass bioethics commission sent out a draft for a new federal bill, dubbed the Dignity of Human Procreation Act, which would subject embryo research and transferring of sperm and egg to federal control.[4]

The Hastings Center argues that America should endorse research on embryos but bring it under federal control with a national authority like England's Human Fertilisation and Embryology Authority or Canada's Royal Commission on New Reproductive Technologies. The center buys into alarmist fears that our present situation is "potentially dangerous" (but leaves this vague) and states that the well-being (it italicizes this phrase for emphasis) of our children, families, and society is at stake.[5]

Just after this report came out, Canada banned payment for surrogate gestation and eggs. Similarly, Italy banned the use of donor sperm, surrogates, or donor eggs in reproductive clinics to create new babies and restricted the use of techniques to "stable heterosexual couples."[6] Everywhere we look, biotechnology is under attack, especially by bioethicists.

PATENTS ON EMBRYOS

Consider the blip on the congressional radar screen at the end of the 2003 legislative session as to whether patents can be issued on human embryos. The issue arose as an amendment that banned patents on "human organisms" to an appropriations bill; it caught everyone by surprise. Worried that it would ban patents on human stem cell lines, the Biotechnology Industry Organization (BIO) lobbied Sen. Sam Brownback (R-KS) to include language that such patents would not be banned, which he did, stating that "nothing in this section shall be construed to affect claims directed to or encompassing cells, tissues, organs, or other bodily components that are not themselves human organisms (including, but not limited to stem cells, stem cell lines, genes, or other living or synthetic organs)."[7] The wording also says that unique processes discovered or created by scientists to create these products could be patented.

Despite the new clarification clause by Senator Brownback, BIO worries that some biological research might be stopped because patents could not be obtained. Right-to-life groups charge that BIO really wants to patent certain kinds of embryos. Chicago Kent law professor Lori Andrews quipped, "If patents on human embryos are allowed, then biotech companies will market babies with certain traits just like Perdue markets chickens." Is that kind of talk really helpful? No thoughtful person believes that. Babies still have to be gestated and raised by humans in families and cannot be marketed "just like Perdue markets chickens."

The truth here is complex. Recall the previous discussion of "what is an embryo"? It is not at all clear at what point modified early human tissue would be eligible for patent and when it would not. That's one problem.

Another is this: suppose you could create a kind of superembryo and make it into a little engine for producing human stem cell lines or twinned embryos. Usually embryos cannot be used for long either way, but suppose you did something to the embryo to change it or stumbled onto a rare genetic embryo that occurred naturally through sexual reproduction, like finding a superior bull or stallion in the animal industry. If you discovered such a superembryo, or created one, you would have something valuable to you, to scientists, and to humanity because lots of research could be furthered by using it.

But you would want to protect your superembryo with a patent. Could it be patented now? The language of congressional bills is contradictory: processes creating new stuff can be patented, but human embryos themselves cannot. So what about a human embryo that is the engine of the process?

Enter the gathering darkness. Obviously, if you're going to the barricades to protect embryos, you are going to fight to the death to prevent embryos from being patented. On the other hand, if an embryo is just a useful dot of tissue that can make special things if the right processes fall in place, then it needs patent protection.

RELIGION AND BIOTECH: TWO WORLDVIEWS

It would be dishonest to pretend that what is at stake here is not a battle between worldviews in conflict. To make that clear, I want to quote from R. Albert Mohler, professor of Christian theology and president of Southern Baptist Theological Seminary in Louisville, Kentucky, who lays out a series of statements that are intended to be reductio ad absurdums but which I take to be the truth.

Professor Mohler first emphasizes that the Christian worldview tells Christians they are made in God's image, and thus tells Christians "who they are." In contrast, "the naturalistic understanding of humanity central to modernity accepts no theistic referent of value. Human beings are cosmic accidents, the fortuitous by-products of blind evolutionary process."[8]

Absolutely correct. We're lucky to be here, given the sterility of the universe as we presently know it. Although other intelligent life is probably out there, right now we don't have any proof of it. The emergence of human consciousness is a billion-to-one shot, maybe trillions-to-one, and we're lucky to have our brief window on the universe while we scoot through our lives.

Mohler then claims that secular ethics must entail eugenics, especially a eugenics driven by "consumer choice," which is his pejorative phrase for the possibility that parents could be given choices over future traits of children.

> More fundamentally, the eugenicist vision represents our human attempt to define ourselves and our destiny. By unlocking the genetic code, by laying naked the genome, we will become masters of our destiny. As human beings, we will define ourselves, improve ourselves, customize ourselves, replicate ourselves, and, in the final act of hubris, redeem ourselves through genetically enhanced and clonally produced progeny.[9]

Keep in mind that Mohler is emphasizing here that the Christian/theistic worldview opposes all the above developments, including the use of the biosciences to "improve ourselves" and to be "masters of our destiny."

Professor Mohler admirably lays out his metaphysical premises, unlike many theists in naturalistic clothing (e.g., Leon Kass and Daniel Callahan). After stating that the heterosexual family unit is divinely ordained, theologian Mohler implies correctly that the naturalistic/secular worldview approves of other ways of constructing families and all novel ways of originating children to make them.

> Given such a worldview, which denies both the Creator and creation, the aspiration to become masters of our own destiny is natural and rational. If we are not created in the image of God, then we will be our own gods. If there is no divine Creator, no Maker of heaven and earth, then we will have to take creation into our own hands. The eugenic temptation is so powerful that only the Christian worldview can restrain it. Scripture alone reveals our creaturely identity, our sinfulness, and the limits of our authority and responsibility. We are not the Creator, and the responsibility to assume control of the universe is not ours. God the Creator rules over all and has revealed his intention for us in laws and commandments that demand our obedience and in limitations that demand our respect.
>
> [In contrast] the very notion of moral limits is foreign to the secular mind.[10]

The beauty of Professor Mohler's presentation is that he lays bare the real assumptions behind many people's reluctance to accept biotechnology, even though they may not believe in God and may reject most of Christian theology for some broader theism. Nevertheless, their reluctance to accept new forms of biotechnology ultimately stems from intuitions shaped from the premises laid out by Mohler.

So, yes, the choices are to accept or reject control of our fate, use the new tools, give people more choices, and reject all theistic limits. Reject the idea that ancient texts written on papyrus can inform our choices today about new forms of life.

Liberal Christians and pro-choice theists will not like Professor Mohler's message and will probably ignore it, seeking some compromise. But I believe Mohler is correct: either we believe that a God exists and sets limits or we reject it in favor of the idea that humans determine their own limits and their own destiny.

Nor is Mohler's message an isolated one. In a recent anthology of religious responses to cloning, each one of the articles unequivocally condemned reproductive cloning as against the natural order, against the family, against God's plan, and bad for society, and almost all condemned embryonic cloning in research.[11]

What scares Christians and other traditional theists is that the new biotech, from cloning to GM veggies to stem cells, is not just science fiction but reality-based expectation. What scares them is that millions, maybe billions, of dollars are being invested in biotech research in hopes of better tools to cure disease, extend life, and enhance our abilities, and people don't invest this kind of money without realistic hopes of a return.

Hence it was predictable that Dan Callahan and Leon Kass would turn their eye on this optimistic kind of research.[12] It is focused on telomerase and telomeres in hopes of curing aging at the cellular level, research that hopes to safely enhance abilities genetically by modifying embryos (not "Baby Einstein," the recorded program on tape, but "Geno Einstein," the congenital enhancement), and research called regenerative medicine that hopes to reverse, retard, or prevent the typical diseases of aging.

What is depressing is that the jihad launched against biotech by the George W. Bush administration, the Kass bioethics council, the Catholic Church, and Protestant fundamentalists is likely to succeed. If we study the rare times in history when science, medicine, and mankind really leaped forward, such as the eighteenth-century Scottish Enlightenment that produced Robert Watts and his steam engine, the historian and philosopher David Hume, and the chemist Robert Boyle, we see that many different cultural, technological, and scientific factors must come together at just the right time to sustain each other and produce an exponential leap. Such a rare serendipity is easily derailed, easily squashed.

The one great hope is that the bio-Luddites will unite the opposition, perhaps under the banner of transhumanism, and use such opposition to motivate us to new heights. Sometimes people need a common enemy to unite against. Perhaps the neocon wave sweeping America and sweeping away biotech can inspire a new pro-choice, pro-technology movement to oppose and transcend it. If not, then some of us must keep a candle lit through the gathering darkness, like the monks who preserved the works of Aristotle through the Dark Ages, preserving the best of our time until the next Enlightenment.

Notes

CHAPTER 1

1. R. C. Lewontin, "The Confusion over Cloning," *New York Review of Books*, 1997; reprinted in Gregory Pence, *Flesh of My Flesh: The Ethics of Cloning Humans* (Lanham, Md.: Rowman & Littlefield, 1998), 129–40; Richard Dawkins, "What's Wrong with Cloning?" originally appeared in two London newspapers in 1997, the *Evening Standard* and the *Independent*, reprinted in *Clones and Clones: Facts and Fantasies about Human Cloning*, ed. Martha C. Nussbaum and Cass R. Sunstein (New York: Norton, 1998), 54–66; Gregory Pence, *Who's Afraid of Human Cloning?* (Lanham, Md.: Rowman & Littlefield, 1998).

2. Lisa Tuttle, "World of Strangers," in *Clones and Clones*, 297–309.

3. Daniel Kevles, *In the Name of Eugenics: Genetics and the Uses of Human Heredity* (New York: Knopf, 1985).

4. C. K. Williams, "My Clone," in *Clones and Clones*, 332–37.

5. Peter Kramer, *Listening to Prozac* (Collingwood, Pa.: DIANE, 1994).

6. Lee Silver, *Re-Making Eden* (New York: Avon, 1998), 249.

7. Robert Wachbroit, "Genetic Encores: The Ethics of Human Cloning," *Report from the Institute for Philosophy & Public Policy*, Fall 1997, 2.

8. Lewontin, "Confusion over Cloning."

CHAPTER 2

1. Adam Phillips, "Sameness Is All," in *Clones and Clones: Facts and Fantasies about Human Cloning*, ed. Martha C. Nussbaum and Cass R. Sunstein (New York: Norton, 1998), 94.

2. Ronald Nakosone, "Buddhist Perspectives on Human Cloning," in *Ethical Issues in Human Cloning: Cross-Disciplinary Perspectives*, ed. Michael Brannigan (New York: Seven Bridges Press, 2001), 96.

3. Richard A. McCormick, "Should We Clone Humans?" *Christian Century Foundation*, reprinted in *Ethical Issues in Human Cloning*, 75.

CHAPTER 3

1. A. Wilcox et al., "Incidence of Early Loss of Pregnancy," *New England Journal of Medicine*, July 28, 1988, 189–94. See also J. Grudzinskas and A. Nysenbaum, "Failure of Human Pregnancy after Implantation," *Annals of New York Academy of Sciences* 442 (1985): 39–44; J. Muller et al., "Fetal Loss after Implantation," *Lancet* 2 (1980): 554–56; "And Then There Was One," *New Scientist*, October 20, 2001, 39–41.

2. Maggie Fox, "Cloned Pigs Open Door for Transplant Organ Farms," January 3, 2002, http://abcnews.go.com/sections/scitech/DailyNews/clonedpigsstudy 020103.html.

3. William Hathaway, "Calf Cloned from Adult Cow Born," *Birmingham News*, June 11, 1998, A5.

4. Patricia Meisol, "Prize Holstein Cloned, Lives on in Daughters," *Baltimore Sun*, January 11, 2002, D1.

5. Mary Yamaguchi, "Japan Sees Cloning as Way to Revive Sick Cattle Industry," *Birmingham News*, November 9, 1998, B3.

6. Hihoko Goto, "That's a Lot of Bull: Clone of a Clone Born," Associated Press, *Birmingham News*, January 25, 2000, A5.

7. Justin Gillis, "Cloned Cows Are Fetching Big Bucks," *Washington Post*, March 25, 2001, A1.

8. Gillis, "Cloned Cows."

9. Peter Fritsch and Jose de Cordoba, "Udderly Fantastic: Castro Hopes to Clone a Famous Milk Cow," *Wall Street Journal*, May 21, 2002, A1.

10. Chuck Oxley, "University Tried for Years to Get Cloning Right," *Casper Star Tribune,* October 19, 2003, 13.

11. Rick Weiss, "First Cloned Horse Created in Italy," *Washington Post,* August 7, 2003, A1.

12. Cesare Galli, quoted by Shaoni Bhattacharya, "World First Cloned Horse Born," *New Scientist,* August 18, 2003, 424, 635.

13. Stephen Jay Gould initially claimed this. See Gould, "Dolly's Fashion and Louis's Passion," *Natural History,* June 1997; reprinted in *Flesh of My Flesh: The Ethics of Cloning Humans,* ed. Gregory Pence (Lanham, Md.: Rowman & Littlefield, 1998), 101–10.

14. Sara Abdullah, "The Living End," *Nature Science Update,* May 27, 1999, www .nature.com/nsu/990527/990527-1.html.

15. T. Wakayama et al., "Aging: Cloning of Mice to Six Generations," *Nature,* September 2000, 318–19.

16. Alan Colman, quoted by Alexa Olesen, Associated Press, February 15, 2003, www.cbsnews.com/stories/2003/02/14/tech/main540689.html.

17. Guatam Naik, "Dolly's Arthritis Fuels Concerns of Health Woes Tied to Cloning," *Wall Street Journal,* January 7, 2003, A1.

18. Andrew Pollack, "In Initial Finding, F.D.A. Calls Cloned Animals Safe as Food," *New York Times,* October 31, 2003, A1.

19. Justin Gillis, "Ailing Dolly, First Cloned Animal, Is Euthanized," *Washington Post,* February 15, 2003, A2.

20. Gina Kolata, "First Mammal Clone Dies; Dolly Made Science History," *New York Times,* February 15, 2003, A4.

21. Jim Erickson, "Cloned Calves Spawn High Hopes, Serious Criticism," *Birmingham Post-Herald,* January 18, 2002.

22. Private communication, Sixth Annual National Undergraduate Conference on Bioethics, Texas A&M University, March 23, 2003.

23. Mark Westhusin, quoted in Gina Kolata, "Animals Cloned for Food No Longer Draw Collective Yawn," *New York Times,* November 4, 2003, D1.

24. Justin Gillis, "Cloned Food Products Near Reality," *Washington Post,* September 16, 2002, A1.

25. See Gregory Pence, *Designer Food: Mutant Harvest or Breadbasket of the World?* (Lanham, Md.: Rowman & Littlefield, 2002).

26. "Organic Trade Association Calls on FDA to Base Animal Cloning Policy on Precautionary Principles," press release, Organic Trade Association, November 3, 2003, www.theorganicreport.org.

27. Denise Grady, "Produce Items Are Vulnerable to Biological Contamination," *New York Times,* November 18, 2003, A3.

28. Melody Petersen and Christopher Drew, "As Inspectors, Some Meatpackers Fall Short," *New York Times,* October 10, 2003, A1, A21. On the safety of eating meat, see also Pence, *Designer Food,* 96–102.

29. Rick Weiss, "At Stake on Your Table," *Washington Post,* November 9, 2003, B1, B5.

30. David Ropeik and George Gray, *Risk: A Practical Guide for Deciding What's Really Safe and What's Really Dangerous in the World around You* (New York: Houghton Mifflin, 2002), 96–121.

31. Ropeik and Gray, *Risk,* 435.

32. David Ropeik and Nigel Holmes, "Never Bitten, Twice Shy: The Real Dangers of Summer," *New York Times,* August 9, 2003, A23.

33. Elizabeth Olson, "Panel Doubts Finding on Cloned-Food Safety," *New York Times,* November 5, 2003, A20.

CHAPTER 4

1. Rick Weiss, "Bouncing Banteng Born to Iowa Cow," *Washington Post,* April 16, 2003.

2. Associated Press, "Reviving a Frozen Zoo," October 14, 2002. Reprinted at www.msnbc.com/news/821198.asp.

3. The Audubon Center's web page is www.auduboninstitute.org/rcenter.

4. Associated Press, "Researchers: Cloning of Endangered Cat a Breakthrough," *Times-Picayune* (New Orleans), November 11, 2003, A1.

5. Agbionet, July 3, 2003, www.agbionet.com.

6. Thomas Fields-Meyer, "Send in the Clones," *People,* September 8, 2003, 90.

7. See www.missyplicity.com.

8. Kristen Hays, "Clone of Cat Not the Same as Original," Associated Press, January 22, 2003, www.cbsnews.com/stories/2003/01/21/tech/main537380.html.

9. Karen Brooks, "Cloned Deer Another First for A&M Team," *Star-Telegram* (Dallas), December 23, 2003.

10. Andrew Pollack, "Cloning a Cat to End the Sniffling and Sneezing of Its Owner," *New York Times*, June 28, 2001.

11. Helen Rumbelow, "Dog's Owners Are Throwing 'Missy' a Clone," *Washington Post*, July 27, 2002, A3.

12. Michael W. Fox, quoted in Charles Graeber, "How Much Is That Doggy in the Window?" *Wired*, March 2000, 22.

CHAPTER 5

1. Aaron Zitner, "Clones, Free Love, and UFOs," *Los Angeles Times*, March 12, 2002.

2. Quoted in Nell Boyce, "Clowns or Clones?" *Newsweek*, January 13, 2003, 48.

3. Zitner, "Clones."

4. *Dateline*, January 3, 2003.

5. Brenda Branswell, "Raelians Had Chortle over Cloning," *The Gazette*, October 9, 2003. Also at www.Canada.com.news.

6. Jack Nichols, "Openly Gay Pro-Human Cloning Advocate Suspects a Hoax," *Gay Today*, December 30, 2002.

7. Zitner, "Clones."

8. Jerry Alder, "Spaced Out," *Newsweek*, January 13, 2003, 50.

9. Boyce, "Clowns or Clones?" 49.

10. Nancy Gibbs, "Abducting the Cloning Debate," *Time*, January 13, 2003, 46.

11. Alder, "Spaced Out," 50.

12. "Baby's Parents Reject Testing, Clonaid Says," *USA Today*, January 25, 2003.

13. David M. Rorvik, *In His Image: The Cloning of a Man* (Philadelphia: Lippincott, 1978).

14. For example, David M. Rorvik, *How to Choose the Sex of Your Baby* (New York: Doubleday, 1997).

15. CNN.com, "Korean Cloning: Out of this World," July 25, 2002, www.religion newsblog.com/117-_Korean_cloning_out_of_this_world.html.

16. Gibbs, "Abducting," 48.

17. Sarah-Kate Templeton, "Clone Baby Couple's Family Dream Dashed after Demands for $80,000," *Sunday Herald* (London), October 26, 2003. For Zavos and his talks, see www.zavos.org.

18. Orville Schell, quoted by Kristen Philipkoski, "Clonaid: Birth of a Media Menace?" W*ired News,* February 2, 2003, www.Wired.com/news/technology.

CHAPTER 6

1. President's Council on Bioethics, *Human Cloning and Human Dignity: Report of the President's Council on Bioethics* (New York: Public Affairs, 2002), 111.

2. Leon Kass, *Life, Liberty, and the Defense of Dignity: The Challenge for Bioethics* (San Francisco: Encounter, 2002), 159.

3. Kass, *Life, Liberty, and the Defense of Dignity,* 159.

4. Kass, *Life, Liberty, and the Defense of Dignity,* 159.

5. Bonnie Steinbock, "Cloning Human Beings: Sorting through the Ethical Issues," in *Human Cloning: Science, Ethics, and Public Policy,* ed. Barbara McKinnon (Urbana: University of Illinois Press, 2000), 76.

6. See Jane Brody, "Gay Families Flourish as Acceptance Grows," *New York Times,* July 8, 2003, citing Suzanne M. Johnson and Elizabeth O'Connor, *Gay Baby Boom: The Psychology of Gay Parenthood* (New York: New York University Press, 2002).

CHAPTER 7

1. Nicholas Wade, "In U.S. First, Doctors Use Stem Cells to Repair Heart," *New York Times,* March 7, 2003; Nancy Tochette, "Stem Cell Transplants for the Heart Face Uncertainties," *Genome News Network,* November 26, 2003, www.genomenews network.org/articles/11_03/heart_stem_cells.html.

2. MedLetter Association, "Health after 50," *Johns Hopkins Medical Letter,* November 2003.

3. Janet McConnaughey, "Two Stem-Cell Transplants Double Odds of Survival," *Birmingham News,* December 25, 2003, 7A.

4. National Bioethics Advisory Commission, "The Science and Application of Cloning," *Cloning Human Beings: Report and Recommendations of the National Bioethics Advisory Commission,* Rockville, Md., June 1997, 20.

5. Denise Grady, "Pregnancy Created Using Egg Nucleus of Infertile Woman," *New York Times,* October 14, 2003, A1.

6. Kyla Dunn, "Cloning Trevor," *Atlantic Monthly,* June 2002, 32.

7. Nicholas Wade and Sheryl Gay Stolberg, "Scientists Herald a Versatile Adult Cell," *New York Times,* January 25, 2002, A7.

8. Robert Weinberg, "Of Clones and Clowns," *Atlantic Monthly,* June 2002, 58–59.

9. *BioNews,* October 27, 2003.

10. Emma Ross, "Scientists Try to Grow Fetus Eggs in Lab," Associated Press, June 30, 2003, www.rednova.com/news/stories/2/2003/06/30/story008.html.

11. Daniel Brison and Brian Lieberman, "An Ethical Way to Provide More Embryos for Research," *BioNews,* November 2003, www.bionews.org.uk/commnetary.lasso?storyid=1899.

12. Rick Weiss, "Mouse Stem Cells Grown into Eggs," *Washington Post,* May 2, 2003, A1.

13. "Embryo Made Using Lab-Built Sperm," BBC News, December 12, 2003, http://news.bbc.co.uk/2/hi/health/3307523.html.

14. Rick Weiss, "Stem Cell 'Master Gene' Found," *Washington Post,* May 30, 2003, A1.

15. Weiss, "Stem Cell 'Master Gene,'" A10.

16 Maggie Fox, "Scientists Gene-Engineer First Human Stem Cells," Reuters, February 9, 2003, http://www.insulinfree.org/cells/yes.htm.

17. I acknowledge that the following account depends heavily on the excellent summary in Ronald Green, *The Human Embryos Research Debates: Bioethics in the Vortex of Controversy* (New York: Oxford University Press, 2001), 26–28.

18. Frederick Grinnell, "Defining Embryo Death Would Permit Important Research," *Chronicle of Higher Education,* May 16, 2003, B13.

19. Leon Kass, *Life, Liberty, and the Defense of Dignity: The Challenges for Bioethics* (San Francisco: Encounter, 2002). For a critique of Kass's book, see Onora O'Neill, "Reason and Passion in Bioethics," *Science*, December 20, 2002.

20. Gilbert Ryle, *The Concept of Mind* (Chicago: University of Chicago Press, 1949).

21. Ruth Macklin, "Dignity Is a Useless Concept," editorial, *British Medical Journal*, December 23, 2003.

22. "Holy See Addresses U.N. on Need to Prohibit Human Cloning," *Zenit* News Service, October 27, 2003, http://www.zenit.org/english/visualizza.phtml?sid=43521.

23. "Therapeutic Cloning Assailed as Creation for Sake of Destruction," ZENIT News Service, October 10, 2003, http://zenit.org/english/visualizza.phtml?sid=43517.

24. John Haas, testimony before the U.S. Senate Subcommittee on Health and Public Safety, June 17, 1997; reprinted in *The Human Cloning Debate*, ed. Glenn McGee (Berkeley: Berkeley Hills Books, 2000), 283.

25. Leon Kass, "How One Clone Leads to Another," *New York Times*, January 24, 2003, A25.

26. Kirk Semple, "U.N. to Consider Whether to Ban Some, or All, Forms of Cloning of Human Embryos," *New York Times*, November 3, 2003.

27. Michael E. Ross, "Law War: Attack of the Clone Debate," February 5, 2003, www.msnbc.com/news/854226.asp?cpl=1.

28. David Hoffman et al., "Cryopreserved Embryos in the United States and Their Availability for Research," *Fertility and Sterility*, May 2003, 1063.

29. Brian Lieberman, "Use of In-Vitro Fertilisation Embryos Cryopreserved for 5 Years or More," *Lancet*, October 4, 2000.

30. http://www.snowflakes.org.

31. "Committee Decides 'Therapeutic Cloning' Can Go Ahead," *BioNews*, May 3, 2002, 2, http://www.aclj.org/news/prolife/020228_cloning_britain.asp.

32. Rick Weiss, "Mature Human Embryos Cloned," *Washington Post*, February 12, 2004, A28.

33. Stephen S. Hall, "Specter of Cloning May Prove a Mirage," *New York Times,* February 17, 2004, D2.

34. Hall, "Specter of Cloning."

35. Dan Vergano, "Research Team Clones Human Embryos," *USA Today,* February 13, 2004, 4A.

36. Weiss, "Mature Human Embryos Cloned," A1.

37. Michael D. Lemonick, "How a Team Cloned Human Cells," *Newsweek,* February 23, 2004, 48.

38. Rick Weiss, "Cloned Embryos Could Help Explain Basis for Disease," *Washington Post,* February 23, 2004, A8.

CHAPTER 8

1. Leon Kass, "How One Clone Leads to Another," *New York Times,* January 24, 2003, A24.

2. Nicholas Wade, "Stanford Institute Is to Study Controversial Stem Cell Manipulation," *New York Times,* December 12, 2002, A1.

3. Stanley Hauerwas, quoted by Michael Shermer, "I, Clone," *Scientific American,* April 2003, 38.

4. Human Cloning Policy Institute (HCPI), press release, November 4, 2003, "Human Cloning Policy Institute Spearheads Global Grassroots Effort to Prevent Therapeutic Cloning Ban at the United Nations," www.clonelaw.org/htm/press-110403.htm.

5. Guatam Naik, "Dolly's Arthritis Fuels Concerns of Health Woes Tied to Cloning," *Wall Street Journal,* January 7, 2003; Andrew Pollack, "In Initial Finding, FDA Calls Cloned Animals Safe as Food," *New York Times,* October 31, 2003, A1.

6. Kass, "How One Clone."

7. Leon Kass, *Life, Liberty, and the Defense of Human Dignity* (San Francisco: Encounter, 2002), 173.

8. "Germany Anguished over Embryonic Life in Cloning Debate," Reuters, November 2, 2003, http://www.biomedcentral.com/news/20031107/07.

9. "Organic Valley Blasts FDA over Cloning Support," *Business Journal* (Milwaukee, Wis.), November 2, 2003.

10. "Plan to Make Human Cloning Safe Set Out," *New Scientist*, October 31, 2003.

11. Robert Edwards and Patrick Steptoe, *A Matter of Life: The Story of a Medical Breakthrough* (London: Morrow, 1980).

12. Irwin Arieff, "UN Anti-Cloning Treaty Seen Heading for Collapse," Reuters, October 3, 3002, http://www.planetark.com/dailynewsstory.cfm/newsid/22463/story.htm.

13. See Seale Harris, *Woman's Surgeon: The Life Story of J. Marion Sims* (New York: Macmillan, 1950), 157.

14. See Gregory Pence, *Classic Cases in Medical Ethics: Accounts of the Cases That Have Shaped Medical Ethics*, 4th ed. (New York: McGraw-Hill, 2004), 313.

15. See Gregory Pence, "Re-Creating Motherhood," in *Re-Creating Medicine: Ethical Issues at the Frontiers of Medicine* (Lanham, Md.: Rowman & Littlefield, 2000), 71–74.

16. Mark Eibert, "Clone Wars," *Reason* 30 (1998); "Human Cloning: Myths, Medical Benefits, and Constitutional Rights," *Hastings Law Journal*, July 2002, 1097–116.

17. Denise Grady, "Pregnancy Created Using Egg Nucleus of Infertile Woman," *New York Times*, October 14, 2003, A1.

18. Grady, "Pregnancy Created," A18.

19. Karby Legett, "China Has Tightened Genetics Regulation: Rules Ban Human Cloning; Moves Could Quiet Critics of Freewheeling Research," *Asian Wall Street Journal*, October 13, 2003.

CHAPTER 9

1. R. Albert Mohler, "The Brave New World of Cloning: A Christian Worldview Perspective," in *Human Cloning: Religious Responses*, ed. Ronald Cole-Turner (Louisville, Ky.: Westminster/John Knox, 1997), 98–99.

2. Eric Posner and Richard Posner, "The Demand for Human Cloning," in *Clones and Clones: Facts and Fantasies about Human Cloning*, ed. Martha Nussbaum and Cass Sunstein (New York: W. W. Norton, 1998), 233–39.

3. The information on the history of eugenics here and below is from Daniel Kevles, *In the Name of Eugenics* (New York: Knopf, 1985).

4. Peter Singer, *One World* (New Haven: Yale University Press, 2003), 1–2.

5. John Robertson, "Two Models of Human Cloning," *Hofstra Law Review*, Spring 1999.

CHAPTER 10

1. Statement by Judy Norsigian, Senate Health, Education, Labor, and Pensions Committee, March 5, 2002, www.ourbodiesourselves.org/clone4.htm.

2. Jeff Opdyke, "Adoption's New Geography," *Wall Street Journal*, October 14, 2003, D1.

3. Ramesh Ponnuru, "Lapse of Reason," *National Review*, February 11, 2002. I am indebted to N. Scott Arnold for pointing out this article to me.

4. Brad Rodu, *For Smokers Only: How Smokeless Tobacco Can Change Your Life* (Los Angeles: Sumner Books, 1998).

5. Bill McKibben, "Unlikely Allies against Cloning," *New York Times*, March 27, 2002, A23.

6. Mary Midgley, *Evolution as a Religion* (London: Routledge, 2002).

7. Michael Crichton, remarks to the Commonwealth Club, San Francisco, September 15, 2003, www.crichton-official.com/speeches/speeches_quote05.html.

8. I am indebted to my colleague N. Scott Arnold for comments on this chapter.

CHAPTER 11

1. "Group Asks World Court to Rule Out Human Cloning," Reuters, October 17, 2003, http://cmbi.bjmu.edu.cn/news/0310/43.htm.

2. Gregory Pence, *Who's Afraid of Human Cloning?* (Lanham, Md.: Rowman & Littlefield, 1998).

3. Louise Vandelac and Marie-Helen Bacon, "Will We Be Taught Ethics by Our Clones? The Mutations of the Living, from Endocrine Disruptors to Genetics," *Bailliere's Clinical Obstetrics and Gynecology* 13 (1999).

4. Quoted in Richard Zaner, "Finessing Nature," *Philosophy and Public Affairs Quarterly*, Summer 2003, 18.

5. Abigail Trafford, "A Sad Case for Cloning," *Washington Post,* June 4, 2002, F1.

6. Julia Sommerfeld, "Coveting a Clone," MSNBC, August 13, 2002, www.msnbc .com/news/768363.asp.

7. John Rawls, *A Theory of Justice* (Cambridge: Harvard University Press, 1973), 108.

8. Gregory Stock, quoted by David Hamilton, "In the Debate on Cloning Humans, UCLA Professor Is One of a Kind," *Wall Street Journal,* June 13, 2002, A1.

9. Malcolm Ritter, "Scientist Says Next Worldwide Flu Outbreak Will Be 'Horrific,'" *Birmingham News*, December 14, 2003, A13.

10. Alan F. Guttmacher, *The Case for Legalized Abortion and Women's Health* (Berkeley, Calif.: Diablo, 1977), 15–17.

CHAPTER 12

1. Linda Carroll, "Alcohol's Toll on Fetuses: Even Worse Than Thought," *New York Times,* November 4, 2003, D1.

CHAPTER 13

1. Laurence Tribe, "Second Thoughts on Cloning," *New York Times,* December 5, 1997, A24.

2. Nick Bostrom, "Human Reproductive Cloning from the Perspective of the Future," www.ReproductiveCloning.net.

3. John Robertson, "Liberty, Identity, and Human Cloning," *Texas Law Review,* May 8, 1999, 1371–456; "Two Models of Human Cloning," *Hofstra Law Review,* Spring 1999, 609–38; Elizabeth Price Foley, "The Constitutional Implications of Human Cloning," *Arizona Law Review,* Fall 2000, 647–730; David Orentlicher, "Cloning and the Preservation of Family Dignity," *Louisiana Law Review,* September 1999, 1019–40; Robert Moffat, "Cloning Freedom: Criminalization or Empowerment in Reproductive Policy?" *Valparaiso Law Review*, Spring 1998, 582–605; Cass Sunstein, "Is There a Constitutional Right to Clone?" Chicago: Public Law and Legal Theory Working Paper no. 22, Social Science Research

Network Electronic Paper Collection, www.papers.ssrn.com/paper .taf?abstract_id=304484; John Charles Kunich, *The Naked Clone: How Cloning Bans Threaten Our Personal Rights* (New York: Praeger, 2003); Mark Eibert, "Human Cloning: Myths, Medical Benefits, and Constitutional Rights," *Hastings Law Journal,* July 2002, 1097–116.

4. *Lifchez v. Hartigan,* 735 F. Sup. 1361 (N.D. Ill., 1990), aff'd, 914 F.2d 260 (7th Cir. 1990), cert. Denied, 111 S. Ct 787 (1991); quoted in "Human Cloning and Genetic Engineering: The Case for Proceeding Cautiously," *Albany Law Review* 65, no. 3 (2003): 649–78.

5. *Eisenstadt v. Baird,* 405 U.S. 438 (1971). See Mark Eibert, "Human Cloning: Myths, Medical Benefits, and Constitutional Rights," in *Ethical Issues in Modern Medicine,* 6th ed., ed. B. Steinbock, John Arras, and Alex John (London: McGraw-Hill, 2003), 656–64.

6. *Skinner v. Oklahoma,* 316 U.S. 535 (1942).

7. Robertson, "Liberty, Identity, and Human Cloning."

8. Richard Doerflinger and Ramon Gonzalez, "Human Cloning: Closer Than You Think," *Western Catholic Reporter,* November 17, 2003, quoting a speech by Doerflinger to the National Pro-Life Conference at Fantasyland Hotel, Edmonton, November 7.

9. Cindy Brovsky, "Trial Begins in Case of Blind Woman Denied Treatment," *Denver Post,* November 10, 2003.

10. "Fertility Clinic Cleared in Blind Woman's Suit," *Birmingham Post-Herald,* November 22, 2003, 5A.

11. Antonio Regaldo, "In Vitro Fertility Proposals by Bush Council Stir Controversy," *Wall Street Journal,* November 21, 2003, B1.

CHAPTER 14

1. J. B. S. Haldane, *Daedalus, or Science and the Future* (London: Dutton, 1924), 44–50.

2. Quoted from Daniel J. Kevles, *In the Name of Eugenics: Genetics and the Uses of Human Heredity* (New York: Knopf, 1985), 185.

3. Scott Rae, "The Advent of the Artificial Womb: A Prospect to be Welcomed?" www.cbhd.org/resources/reproductive/rae_2003-01-29.htm.

4. This phrase was suggested to me by Tom Tomlinson.

5. H. C. Liu, Z. Y. He, C. Mele, M. Damario, O. Davis, and Z. Rosenwaks, "Hormonal Regulation of Expression of Messenger RNA Encoding Insulin-Like Growth Factor Binding Proteins in Human Endometrial Stromal Cells Cultured In Vitro," *Molecular Human Reproduction* 3, no. 1 (1997): 21–26; Robin McKie, "Men Redundant? Now We Don't Need Women Either," *The Observer* (London), February 10, 2002.

6. McKie, "Men Redundant?"; Ronald Bailey, "Babies in a Bottle," *Reason Online*, August 20, 2003, reason.com/rb/rb082003.html.

7. "Japanese Scientist Develops Artificial Womb," Reuters, July 18, 1997, http://www.w-cpc.org/news/reuter7-97.html.

8. Jonathan Knight, "An Out of Body Experience," *Nature*, September 2002, 106–7.

9. Rae, "Advent."

10. Rae, "Advent."

11. Helena Ragone, *Surrogate Motherhood: Conception from the Heart* (Boulder: Westview, 1994), 4–6.

12. Nancy Reame, "Surrogate Mothers Feel Disappointment," speech presented at the Sixteenth World Congress on Fertility and Sterility, American Society of Reproductive Medicine Annual Meeting, San Francisco, October 5, 1988.

13. Paul Ramsey, *The Fetus as Person* (New Haven: Yale University Press, 1976).

CHAPTER 15

1. *Humanzee,* the Discovery Channel, June 7, 2003; see also www.rotten.com/library/cryptozoology/humanzee.

2. Rick Weiss, "Cloning Yields Human–Rabbit Hybrid Embryo," *Washington Post,* August 14, 2003, A4.

3. Andrew Stern, "Artist Seeks to Free His Glowing Creation—Rabbit," Reuters, September 23, 2000, http://www.ekac.org/reuters.html.

4. Andrew Pollack, "Gene Altering Revolution Is about to Reach the Local Pet Store: Glow-in-the-Dark Fish," *New York Times,* November 22, 2003, A11.

5. James Gorman, "When Fish Flouresce, Can Teenagers Be Far Behind?" *New York Times*, December 2, 2003, D3.

6. Bernard Rollin, *Animal Rights and Human Morality* (Buffalo, N.Y.: Prometheus, 1992).

7. Charles Darwin, *The Origin of the Species by Natural Selection* (London: John Murray, 1859), 52. Facsimile of the first edition reprinted by Harvard University Press (1964). Quoted from Rachels, *Created from Animals* (New York: Oxford University Press, 1990), 195.

8. Jason Roberts and Françoise Baylis, "Crossing Species Boundaries," *American Journal of Bioethics*, Summer 2003, 3.

9. Richard Lewontin, "The Dream of the Human Genome," *New York Review of Books*, May 28, 1992, 31–40.

10. David Castle, "Hopes against Hopeful Monsters," *American Journal of Bioethics* (Summer 2003): 3, 28–30.

11. R. Albert Mohler, "The Brave New World of Cloning: A Christian Worldview Perspective," in *Human Cloning: Religious Responses*, ed. Ronald Cole-Turner (Louisville, Ky.: Westminster/John Knox, 1997), 92–93.

12. Roberts and Baylis, "Crossing Species Boundaries," 2.

13. Peter Singer, *Animal Liberation* (New York: New York Review of Books, 1975); Rollin, *Animal Rights*; Bernard Rollin, *The Frankenstein Syndrome: Ethical and Social Issues in the Genetic Engineering of Animals* (New York: Cambridge University Press, 1995).

14. Hilary Bok, "What's Wrong with Confusion?" *American Journal of Bioethics*, Summer 2002, 25.

15. John Charles Kunich, *The Naked Clone: How Cloning Bans Threaten Our Personal Rights* (New York: Praeger, 2003).

16. Andrew Pollack, "Cloning a Cat to End the Sniffling and Sneezing of Its Owner," *New York Times*, June 27, 2001; "More on Allergy Free Cats," *Sinus News*, July 15, 2001.

17. Carol Smith, "Supplies of Arthritis Drug Running Low," *Seattle Post-Intelligencer Reporter*, April 24, 2002.

18. Andrew Wyatt, "Eye on America," *CBS Evening News*, August 6, 2001.

19. Interview by Andrew Wyatt, "Eye on America," *CBS Evening News,* August 6, 2001.

20. James Rachels, *Created from Animals: The Moral Implications of Darwinism* (New York: Oxford University Press, 1990).

21. Singer, *Animal Liberation.*

22. Rachels, *Created from Animals,* 129.

23. Clive D. L. Wynne, "'Willy' Didn't Yearn to Be Free," *New York Times,* December 27, 2003, A24.

CHAPTER 16

1. Alvin F. Poussaint, "The Hope of Head Start," *New York Times,* July 22, 2003, A23.

2. Richard Hernnstein and Charles Murray, *The Bell Curve: Intelligence and Class Structure in American Life* (New York: Free Press, 1999).

3. Charles Murray, *Losing Ground* (New York: Basic, 1984).

4. Judith Rich Harris, *The Nurture Assumption: Why Children Turn Out the Way They Do* (New York: Free Press, 1998).

5. David Reiss, quoted by Sharon Begley, "The Nature of Nurturing," *Newsweek,* March 27, 2000.

6. Lee Silver, *Re-Making Eden: Cloning and Beyond in a Brave New World* (New York: Avon, 1997), 246–47.

7. Silver, *Re-Making Eden,* 247.

8. Amy Chua, *World on Fire: How Exporting Free Market Democracy Breeds Ethnic Hatred and Global Instability* (New York: Doubleday, 2003).

9. Chua, *World on Fire,* 3–4.

10. Loren Lomasky, "Defending Tolerance: Values, Liberty, and Democracy," *Reason,* November 2003, 53.

11. Chua, *World on Fire,* 6.

12. Marc Lacey, "Once Outcasts, Asians Again Drive Uganda's Economy," *New York Times,* August 17, 2003.

13. John Rawls, *A Theory of Justice* (Cambridge, Mass: Harvard University Press, 1971), 511.

14. Wayne Arnold, "Singapore Goes for Biotech: A $280 Million Complex, Complete with Mice," *New York Times*, August 26, 2003.

15. This point about the Special Olympics came from a short talk entitled "Human Variation and Fairness in Sports and Games," by Leslie Francis at the Transvision 2003 Conference at Yale University on June 28, 2003.

CHAPTER 17

1. Wendy Goldman Rohm, "Seven Days of Creation," *Wired*, January 2004, 122–29.

2. Kyla Dunn, "Cloning Trevor," *Atlantic Monthly*, June 2002.

3. Erik Parens and Lori Knowles, "Reprogenetics and Public Policy," *Hastings Center Report*, supplement, July-August 2003, S1–S24.

4. Staff Working Paper, "Biotechnology and Public Policy: Biotechnologies Touching the Beginnings of Life: Defending the Dignity of Human Procreation," www.bioethics.gov/background/bpp_defend_dig.html.

5. "Biotechnology and Public Policy," S3.

6. "Italy Bans Donor Sperm and Eggs," Reuters, December 11, 2003, http://news.bbc.co.uk/1/hi/world/europe/3311031.htm.

7. Rick Weiss, "Funding Bill Gets Clause on Embryo Patents," *Washington Post*, November 17, 2003, A4.

8. R. Albert Mohler, "The Brave New World of Cloning: A Christian Worldview Perspective," in *Human Cloning: Religious Responses*, ed. Ronald Cole-Turner (Louisville, Ky.: Westminster/John Knox, 1997), 96.

9. Mohler, "Brave New World," 99.

10. Mohler, "Brave New World," 102.

11. Ronald Cole-Turner, ed., *Human Cloning: Religious Responses* (Louisville, Ky.: Westminster/John Knox, 1997).

12. Daniel Callahan, *What Kind of Life? The Limits of Medical Progress* (New York: Simon & Schuster, 1990); Leon R. Kass, "L'Chaim and Its Limits: Why Not Immortality?" *First Things*, May 2001, 17–24.

Index

About the Author

Gregory Pence has taught for nearly thirty years in the Philosophy Department and School of Medicine at the University of Alabama at Birmingham. He testified about cloning before committees of the U.S. Congress and California Senate, and he has appeared on *The Point* with Greta Van Susteren on CNN, *The Early Show with Bryant Gumbel* on CBS, *CNN Late Edition with Wolf Blitzer*, and National Public Radio. He has also been interviewed by *Time* magazine, the *New York Times*, and most national publications and has published in *Newsweek*, the *New York Times*, and the *Wall Street Journal*. He has given invited talks at more than two hundred universities.

Pence's book *Designer Food: Mutant Harvest or Breadbasket of the World?* (Rowman & Littlefield) was named a Choice Outstanding Academic Title in 2003. His *Classic Cases in Medical Ethics: Accounts of the Cases That Shaped Medical Ethics* (McGraw-Hill, 4th ed., 2004) is one of the standard textbooks of bioethics.

He grew up in Washington, D.C., was graduated from the College of William and Mary, and earned his doctorate from New York University in 1974.